The Sacred Balance

David Suzuki

with **Amanda McConnell** | **Adrienne Mason** | **Ian Hanington**

Foreword by **ROBIN WALL KIMMERER**

Afterword by **BILL MCKIBBEN**

The Sacred Balance

Rediscovering Our Place in Nature

25TH ANNIVERSARY EDITION

DAVID SUZUKI INSTITUTE

GREYSTONE BOOKS
Vancouver/Berkeley/London

23 24 25 26 27   6 5 4 3 2

Greystone Books Ltd.
greystonebooks.com

David Suzuki Institute
davidsuzukiinstitute.org

Cataloguing data available from Library and Archives Canada
ISBN 978-1-77164-986-5 (pbk)
ISBN 978-1-77164-987-2 (epub)

Editing for original edition by Nancy Flight
Copy editing for revised edition by James Penco
Proofreading for revised edition by Meg Yamamoto

Cover and text design by Jessica Sullivan
Front cover illustration by Dóra Kisteleki
Illustrations by Vicky Earle

Printed and bound in Canada on FSC® certified paper. The FSC®
label on this paper ensures responsible use of the world's forest resources.

Greystone Books thanks the Canada Council for the Arts, the British Columbia
Arts Council, the Province of British Columbia through the Book Publishing Tax Credit,
and the Government of Canada for supporting our publishing activities.

Canada

MIX
Paper from
responsible sources
FSC® C103567

BRITISH COLUMBIA

BRITISH COLUMBIA
ARTS COUNCIL
An agency of the Province of British Columbia

Canada Council Conseil des arts
for the Arts   du Canada

Greystone Books gratefully acknowledges the xʷməθkʷəy̓əm (Musqueam),
Sḵwx̱wú7mesh (Squamish), and səl̓ílwətaʔɬ (Tsleil-Waututh) peoples on
whose land our Vancouver head office is located.

*This book is dedicated with love to
Kaoru and Setsu, my parents,
who taught me to love nature and to respect my elders;
Tamiko, Troy, Laura, Severn and Sarika,
my children, who lifted my sight into the future; and
Tara, my wife, partner and best friend,
who showed me the meaning of love and commitment.*

Contents

Foreword

T HE BEANS RATTLE in their pods as I twist them from the vines. Gathering time used to be right after the October killing frost, but it's late November and still it has not come. New bean flowers blossom beside the dry pods in this shifted season. Wildfires blaze, storms thrash, alarming headlines of climate crisis are everywhere—and show up here in my strange untimely harvest. It makes me even more grateful for their gift of food. In my Anishinaabe language, the word for "seed" and the word for "gift" share a common root. I am grateful too for this new edition of *The Sacred Balance*, which is an ongoing gift. A book is its own kind of seed.

David Suzuki was among the first scientists to bring the threats of climate catastrophe to the consciousness of his legions of listeners and readers. The messages were urgent when *The Sacred Balance* was first published in 1997, when climate disruption was just over the horizon and our home was not yet on fire—and they are even more timely today, when the world we love is melting away like its glaciers. In heartbreaking waves of biodiversity loss, the evolutionary marvels of our more-than-human kinfolk vanish forever. His warnings were clear and compelling and yet here we are, twenty-five years later, more in need of their lessons than ever.

I think it is a profound privilege to be a scientist. This path offers the gifts of seeing into the lives of other beings, immersion in appreciation of the physical and ecological laws that govern the unfolding of life on the planet, the connection to mystery.

All too often scientists keep that gift for themselves, communicating only with one another in language that excludes the public. We all pay a price for this practice, in reduced scientific literacy and estrangement from the natural world. If we scientists don't tell the stories of other beings, who will? Their absence from our consciousness can lead us to imagine that *Homo sapiens* is alone in the world, atop some fictional pyramid of human exceptionalism, in charge of an Earth we mistakenly believe belongs to us. *The Sacred Balance* is a strong counterweight to such error.

As I hold these seeds in my hand and contemplate their purpose, I'm reminded of our old teaching that gifts and responsibilities are two sides of the same coin. Sequestered inside these hard seed coats are living beings. Seeds hold the gift of carrying knowledge across the years and the responsibility to use the knowledge encoded in that embryo to multiply themselves to feed the people. Gifts and responsibilities walk hand in hand.

In Western science, knowledge for knowledge's sake is unencumbered by any restrictions on its use, while in many Indigenous societies, knowledge is intimately coupled with responsibility for careful use and respectful sharing. In the pages of *The Sacred Balance*, David Suzuki embraces the responsibility that comes with the gift of being a scientist: to use that knowledge in service of the flourishing of life.

Many scientists draw a strict line between knowledge and advocacy, adhering to the dubious notion that speaking out about the political and social implications of scientific knowledge in some way diminishes their objectivity as researchers.

Fortunately, that wall of separation is eroding with the imperative to have science play its rightful role in policy-making. The lifetime work of David Suzuki, as an activist and scholar, has been to carry the gift of science outside the research lab, into the minds and hearts of citizens. In *The Sacred Balance*, he has polished the art of the science storyteller. He eloquently expresses that the gift of knowledge is not enough; we have the responsibility to act from that knowledge to create a right relationship.

This new edition amplifies the message that we need to restore a sense of the sacred to our relation with the Earth. The power of *The Sacred Balance* rests perhaps in the simple container of its title, which feels like an invitation—to not only comprehend and marvel at the co-evolved ecological balance, but participate in it, to assume our rightful place in the processes that keep life going. *The Sacred Balance* calls us in; it asks us to be a better species, to align our human laws with the laws of nature. Suzuki's prose not only illuminates the scientific workings of the world, but invokes a kinship and purpose that we long for, to belong to something simultaneously as immense as the atmosphere and as intimate as breath. I linger with his words:

> From our first cry announcing our arrival on Earth to our very last sigh at the moment of death, our need for air is absolute. Every breath is a sacrament, an essential ritual. As we imbibe this sacred element, we are physically linked to all of our present biological relatives, countless generations that have preceded us and those that will follow. Our fate is bound to that of the planet by the gaseous exhausts of fires, volcanoes and human-made machines and industry.
>
> Once we have restored the breath of life to its rightful primacy—the first above all other human rights and responsibilities, the reference point from which all decisions flow—we can

start to work in the long term to revive an ancient equilibrium. Using nature as our touchstone, we can play our part once more in life's long collaboration with the air.

The science of anthropogenic impacts on the atmosphere is clear, but the climate crisis is more than an ecological problem or a technical challenge. The roots of the crisis go deeper than bio-geochemistry, beyond governance and economics. What we face is a spiritual crisis, grounded in a flawed worldview that is a hulking barrier to achieving balance. Suzuki writes:

> Through our loss of a worldview, our devotion to consumer-ism and our move into the cities and away from nature, we have lost our connection to the rest of the living planet. We must find a new story, a narrative that includes us in the continuum of Earth's time and space, reminding us of the destiny we share with all the planet's life, restoring purpose and meaning to human existence.

We are much further from achieving this balance than we were in 1997. We know what to do but we don't do it. What is holding us back? The answers are legion, but if I had to name a root cause of our wilful deafness to the suffering of the Earth, it would be the anthropocentric worldview that has powered the losses of colonialism and extractive capitalism and has us teetering on the brink of ecological catastrophe.

This calls us to undertake a transformation from a worldview of domination and exploitation to a kin-centric worldview that acknowledges our utter dependence upon the life-giving gifts of other species, before whom we can only bow in gratitude. Gratitude awakens the impulse to return the gifts, to enter the covenant of reciprocity.

Like seeds passed from grandparent to grandchildren, this book embodies an intergenerational sharing of knowledge. I am heartened by its celebration and solidarity with the youth-led climate justice movement. Seeds carry the story of our past, but also resilience for the future. I wonder if these late-formed seeds will carry new adaptations to the changes that are upon us.

Adaptation, the ability to change in such a way that enhances survival and flourishing, occurs not only in the resilience of an individual, but within populations over time. Accumulated changes that increase fitness in a changing world are encoded in genetic libraries and manifest in altered structures, functions and behaviours. Suzuki suggests that restoring the sacred to our relationship with nature may be the most important adaptation we can make. Here, I think, is the most vital message he carries: that spiritual beliefs and practices, encoded in a worldview, are of immense adaptive value. He writes: "Spirituality may be our chiefest local adaptation—the means by which we touch the sacred, hold together against disintegration. The forms and varieties of spiritual belief and ritual among cultures on Earth may be another example of evolution's incredible, extravagant invention of ways for life to survive."

And these adaptations need not arise by novel mutation. We already live in the presence of the model and the guidance of Indigenous worldviews, which marry science with spirit, gift and responsibility. The very knowledge that colonialism sought to violently erase and replace offers an alternative to the anthropocentric worldview, a pathway toward mutual flourishing of land and people—not to appropriate and take as our own, but to learn from with gratitude and humility, a universe of knowing generated, maintained and safeguarded by Indigenous Peoples.

In this time of planetary crisis, we are awash in information and yet starving for wisdom. While powerful in generating knowledge, Western science—in its separation of values and knowledge, of object and subject, of matter and spirit—is incapable of growing wisdom. David Suzuki has long recognized and been an ally in lifting up another kind of holistic science, the ancient and contemporary complex of knowledge, practice and belief that is Indigenous knowledge—where systematic generation of knowledge is guided by the human values of respect, responsibility, reciprocity and reverence. This new revised edition amplifies Suzuki's long alliance with Indigenous knowledge holders. At a time when 80 per cent of the world's remaining biodiversity is safeguarded by Indigenous Peoples, yet they legally own only 10 per cent of the land mass, this book recognizes the imperative for Western science to learn from Indigenous knowledge.

As I open the pods, the seeds clatter into the bowl, black ovals swirling with magenta and pink. These seeds were handed to me from another Indigenous gardener, just as they were passed to her by a long line of ancestors. Heritage seeds are ancient and new at the same time. What is it that makes one seed variety last while others are lost? Like knowledge, a seed endures when planters care for it, safeguard its genome and plant it again and again, for the vigour of its growth, its beauty, the stories it carries, the way it nourishes the people.

What makes a book last, to be reread and passed to the next generation? *The Sacred Balance* is planted again because it sows a vision of a future we want to live in, with guidance to get there. Like my bean seeds, these pages carry a message between generations that nourishes us in ways both physical and spiritual.

—ROBIN WALL KIMMERER

Introduction to the 25th Anniversary Edition

WHEN RACHEL CARSON'S *Silent Spring* was published in 1962, it galvanized a global environmental movement that grew to millions of people worldwide. Carson opened our eyes to the limitations of "reductionist" science and the need for a more holistic view of ecosystems and their interactions. It was exciting to see environmental concerns finally gain the attention they warranted. I helped celebrate many "victories"—agricultural land protected, wilderness areas set aside as parks and reserves, and dam projects halted, among others. But no matter how many wins we celebrated, new threats arose: protected land nibbled away, mining and logging allowed in parks, halted projects renewed and environmental legislation overturned.

Our celebrations proved to be short-lived, as proponents of the destructive activities simply renewed or revised their proposals, found alternative routes to the same ends or exploited new governments. Our victories were pyrrhic because we failed to address the values, assumptions and beliefs that underlie our destructive demands and activities. The way we see the world shapes the way we treat it. If we regard a mountain as a deity rather than a pile of ore, a forest as a sacred grove rather than lumber and pulp, other species as our biological kin rather than resources, the planet as our mother and life-giver and not an opportunity, then our actions will reflect far greater humility, respect and responsibility.

For almost our entire existence, we were nomadic hunter-gatherers who carried all we possessed as we followed plants and migrating animals through the seasons. We understood our place as a small strand in a complex web of relationships with other plant and animal species and with air, water, soil and sunlight. This "ecocentric" perspective invokes gratitude for nature's productivity and generosity, accompanied by a sense of responsibility to act properly to maintain those gifts.

When it was first published in 1997, *The Sacred Balance* may not have provoked as widespread a reaction as *Silent Spring*, but it's as relevant today as it was then, if not more so—especially in light of the growing recognition that we must reconcile ways of knowing to gain a greater understanding of our place in a world with serious challenges of our own making. By placing too much faith in science and not enough in the knowledge of people who have lived in place, we have lost essential wisdom and understanding.

Human life has been changing at explosive speed. We have shifted from the notion of embeddedness and responsibility to a belief that our great intelligence makes us the most important animal, endowed with the ability to extricate ourselves from the web of relationships to a position of exploitation, control and management. In such an "anthropocentric" worldview, nature is ours for the taking.

Today, our rapidly increasing population, technological capability, consumptive demand and global economy have made us unique among the life forms that have occupied the planet for 4 billion years. We are now altering the physical, chemical and biological properties of the planet on a geological scale. Many scientists refer to our time as the Anthropocene epoch, the Age of Humans. As powerful as we are, we remain too ignorant of how everything within the biosphere interacts to be able to guide our actions properly, and so we end up undermining the planet's life-support systems—air, water, soil, photosynthesis and biodiversity—that are the foundations of our existence and well-being. How did we reach this moment?

Molecular biologists have isolated, sequenced and manipulated the genetic material DNA, enabling them to compare samples between different human populations. The research suggests people moved and separated over time, but all trails lead back to Africa 150,000 to 200,000 years ago.

When we appeared, herds of animals far exceeding today's abundance and diversity roamed the great grasslands of Africa. Among them, humans were neither threatening nor impressive in numbers, size, speed, strength or sensory acuity. We were just upright, furless apes. When we passed, other species likely didn't tremble in fear. Our evolutionary advantage was invisible, the 1.5-kilogram organ in our skulls. Our brains more than compensated for our lack of special physical or sensory qualities, as the fossil record suggests we killed off other species with tools as simple as spears and stone axes.

After forty or fifty millennia, we began to move. Perhaps our numbers had grown to the point where tribal disputes and territorial battles were becoming common, where plants or animals on which we depended were becoming less abundant. Maybe it was partly teenagers searching for adventure or excitement. For whatever reason, we sought new lands and entered them as an alien species. With flightless birds, slow-moving sloths and abundant fish, these "new to us" ecosystems offered opportunities. As an invasive species, we had to learn through experience. Observation, mistakes, successes and failures formed the culture of successive generations in new lands. This became the basis of

Indigenous knowledge worldwide, and the survival of cultures a measure of its success.

Ten millennia ago, domestication of plants and then animals ended our dependence on hunting and gathering. By planting and harvesting edible plants, humans began the deliberate use of nature for a more certain future. The Agricultural Revolution changed the course of history, providing dependable food sources, creating stable communities in place and enabling population growth and elaborations of music, art, language and culture. Within a few millennia, people had domesticated modern forms of grasses (wheat, corn and rice) and animals (dogs, sheep, cows, horses and chickens).

But as populations and villages grew, religions arose to guide human behaviour. Many promoted the idea that we are special and different from the rest of nature. In the Bible's book of Genesis, God tells Adam and Eve, "Be fruitful and multiply, and fill the earth and subdue it; rule over the fish of the sea and the birds of the air and every creature that crawls upon the earth."

Fuelled by agriculture, populations grew from villages into states, nations and empires that rose and fell over centuries. From just another ape on the African grasslands, we saw ourselves created in the image of gods for whom all of existence was ours for the taking. Chiefs, kings and emperors arose, acquiring wealth, power and dreams of immortality.

The Renaissance emerged from the Middle Ages in Europe from the fourteenth to seventeenth centuries. Human thought and creativity flowered in philosophy, art and the new discipline of science. In 1597, Francis Bacon's famous phrase *"scientia potestas est"* ("knowledge is power") expressed his belief in the importance of scientific inquiry. Well schooled in ancient thinkers like Aristotle, he sought to modernize inquiry through organization of facts, testing ideas and experimentation. He is regarded as the father of modern science.

In 1637, René Descartes famously wrote, "*Cogito, ergo sum*" ("I think, therefore I am"), positing that the very act of thinking confirmed the reality of one's existence.

Mathematician and physicist Isaac Newton invented calculus and gave measured precision to descriptions of the motion of heavenly bodies, while also studying alchemy and the Bible. His work seemed to validate the notion of the cosmos as an immense clockwork mechanism that could be understood by examining its parts in detail.

The ideas of these giants in philosophy and science continue to reverberate. If the properties expressed at every level of organization—from subatomic particles up to populations, planets and galaxies—are simply the sum of everything at each level, then each part of an object or system can be analyzed in detail and ultimately assembled with the other parts to explain the properties of the whole. This reductionism leads to the conclusion that the behaviour of a cell, organ or whole organism is simply the consequence of its full atomic composition. This is the basis of much of research in modern physics, chemistry and biology, even though physics has long shown this assumption is wrong, and that "emergent" properties result from the combination of elements—the whole being greater than the sum of its parts. Perhaps the best illustration is water (H_2O), whose properties cannot be predicted from the properties of atomic hydrogen and oxygen.

Three centuries ago, the Industrial Revolution heralded support for Baconian optimism about the human triumph over nature, and thus our escape from the tyranny of natural law. We had harnessed the nutrition of domesticated plants and animals and expanded our limited strength and endurance by use of other animals like horses and oxen. Gunpowder weapons lifted us out of the array of claws, fangs and shells. Eventually, machines could perform jobs beyond biological constraints, working tirelessly for long stretches, performing repetitive tasks and producing flawless, identical products.

Technology, created by and for humans, enabled us to travel faster, farther and with little effort, now even outrunning the speed of sound. We are air-breathing terrestrial animals, yet our intelligence lets us explore and live underwater, dive to the floors of the deepest oceans, pierce Earth's crust, decapitate mountains, reroute rivers and drain lakes, destroy entire ecosystems and escape gravity to live in airless space.

Believing our great intelligence has elevated us out of nature and natural laws, we think the only limit to our potential is imagination and creativity. We are indeed a clever species. From our beginnings, our intelligence has enabled us to survive and flourish. But when we assume the mantle of dominance, we forsake acknowledging our ignorance and responsibility. The history of contemporary culture illustrates the disastrous consequences.

Over and over, in our exuberance about harnessing a new discovery, such as the release of vast amounts of energy by atomic fission or the chemical inertness of the large molecule CFC, we employ the application only to later discover unpredictable consequences.

We haven't learned enough from the many historical lessons about the enormity of our ignorance. Proposals are underway to continue massive-scale technological innovation in biotechnology (engineering life forms according to DNA manipulation), nanotechnology (creation of molecules that interact with metabolic processes), artificial intelligence (machines that "think" at the speed of light compared with human neurons, which transmit information in metres per second) and geoengineering (combatting climate change by deliberate intervention and manipulation of the biosphere).

Anthropocentrism lies at the heart of the global eco-crisis today because it pervades the systems—legal, economic, political—that determine and guide the way we live. Legal systems define individual rights and property.

Having assumed "ownership" of the entire planet, we delineate and adjudicate within boundaries we draw around countries, provinces, municipalities and private property. Human borders are a strange notion. In 1986, Swedish scientists detected a huge spike in atmospheric radioactivity, with all indications that it originated in what was then the Soviet Union. It was later revealed to have come from a meltdown at the Chernobyl nuclear plant. Human boundaries mean nothing to nature, yet in trying to adjudicate within them, we can never hope to "manage" natural systems, in which everything is interconnected and interdependent.

Laws are for humans, yet they spill over in the way we exploit other species and non-biological elements. But where in our legal system is the right of a songbird to live life as it evolved to do? Where is the right of a river to flow as it has for millennia? What about the right of a forest to exist and flourish as a diverse community of organisms? In our anthropocentrism, we assume we have the right to administer and adjudicate for the rest of existence. And so, when environmentalists attempt to protect ecological elements, it is within the context of the way those elements affect us. The legal game is rigged in favour of humans.

Human creativity and productivity are the basis of economies. Since it is assumed there are no limits to our imagination, economists believe endless growth is possible, and is the very definition of success and progress. Following the Great Depression of the 1930s, the Second World War provided the stimulus for global economic recovery. The transition from the wartime to peacetime economy was built on consumerism. With disposability replacing durability in goods, markets never needed to run out of clients. The notion of "planned obsolescence" built into products like cars has extended into clothing and electronic gadgets. Even the great triumph of the successful development of COVID-19 vaccines led to the production and delivery of countless syringes that were used once and discarded.

As we confront the impossibility of endless economic growth, we should redefine the meaning of prosperity and wealth. Everything we have comes from nature. We need to answer the important questions: What is an economy for? Are there no limits? How much is enough? Are we happier with the flood of consumer items now overwhelming us?

Democracy, whereby all eligible voters help shape government policy, is often considered the best governance system devised. The continuing revision of voter eligibility to include women, people of colour and prison populations is testimony that it is not perfect or infallible but that it can evolve. It hints at problems intrinsic to a political system operating within anthropocentric values. In a democracy, we elect people to represent us but fail to reflect the diversity among different genders, ethnicities, wealth statuses and economic sectors. After a successful election, a politician's highest priority becomes re-election, which means political decisions and actions must be accounted for within the interval between elections. Issues like climate change and biodiversity loss, which have been brewing and will last for decades or centuries, will not be resolved within election cycles. With rare exceptions, principles are subordinated to the service of "party politics." Another major problem of anthropocentric politics is that those most profoundly affected by decisions governments make or don't make—children and future generations—don't vote. And the atmosphere, oceans, forests, soil, water and mountains have no representation except as they relate to humanity. Thus, forest, environment and fisheries and oceans ministers do not act in the interests of forests, the environment or fish and oceans, because only people vote.

It's time to reformulate these legal, economic and political systems that guide human behaviour and activity to ground them within an ecocentric perspective.

The endless conflicts that stall critical action are what make *The Sacred Balance* so relevant and necessary today. In this book, we try to envision a future within the context of our relationship with the planet, respecting and incorporating wisdom and insights passed through generations of Indigenous Peoples worldwide.

It's urgent that we shift from anthropocentrism to ecocentrism. The science is in, the consequences of human activity are clear and the threats to our survival are a stark reality. Failure to acknowledge our continuing embeddedness in nature and utter dependence on its services for our health and survival has led us into repeated destructive ways. We are biological creatures, as dependent on clean air, water, soil, food and sunlight (through photosynthesis) as any other animal.

Many Indigenous Peoples refer to plants and animals as their "relatives" or "kin." They express respect, and when they take the life of an animal to eat, they may apologize for a necessary act and thank their relative for the sacrifice. Such a belief carries a sense of greater care and respect than if an object is simply a "resource" or "opportunity." The ability to sequence the DNA of whole genomes of organisms reveals a surprising corroboration of Indigenous beliefs. We *are* related to all life. The degree of relatedness to an eagle, bear, salmon or cedar tree is proportional to the length of time since we shared a common ancestor. It took most of the 3.9 billion years after life arose for bacteria to evolve all the basic rules and principles of metabolism, cell division and reproduction. Multicellularity and all of the shapes of life were mere recent elaborations on those basic microbial units that were the ancestors of us all.

There is no separation between us and non-biological elements of the biosphere. We don't wall ourselves off with our skin. Air is in us and circulating through our bodies to keep our life force ignited. When air exhausts from our nostrils, it goes into the bodies of plants and animals around us. Water inflates every cell in our body while the atoms and molecules from carcasses of plants and animals we consume form the structures and architecture of our bodies. Every movement we make is set off by metabolic combustion that releases the energy of sunlight to fuel our activity. We are the Earth, created of and animated by its sacred elements: earth, air, fire and water. As social animals, it is love that makes us fully human, and we know who we are through spirit and ceremony.

We must find a sacred resilience and balance within a constantly changing world.

—DAVID SUZUKI

Prologue

S UPPOSE THAT 200,000 years ago, biologists from another galaxy searching for life forms in other parts of the universe had discovered Earth and parked their space vehicle above the Rift Valley in Africa. At the moment of our species' birth, mammoths, sabre-toothed cats, huge moa birds and giant sloths still roamed the planet. Those intergalactic visitors would have gazed upon vast grasslands filled with marvellous plants and animals, including a newly evolved species, *Homo sapiens*.

It is highly unlikely that those alien scientists would have concentrated their attention on this infant upright ape species in anticipation of its meteoric rise to pre-eminence a mere two hundred millennia later. After all, those early humans lived in small family groups that didn't rival the immense herds of wildebeest and antelope. In comparison with many other species, they weren't especially large, fast or strong, or endowed with special sensory acuity. Those early humans possessed a survival trait that was invisible because it was locked within their skulls and only revealed through their behaviour. Their immense and complex brains conferred tremendous intelligence, along with a vast capacity for memory, an insatiable curiosity and an astonishing creativity—abilities that more than compensated for their physical and sensory deficiencies.

That newly evolved human brain invented a novel concept called the "future." In reality, all that exists is the present and our memories of what is past, but by creating the notion of a future, we were unique in recognizing that we could influence events to come by what we do in the present. By looking ahead, we could anticipate potential danger and opportunities. Foresight was the great advantage that catapulted *Homo sapiens* into a position of dominance on the planet.

The late eminent Nobel laureate François Jacob suggested that the human brain is "hardwired" to require order. Chaos is terrifying to us, because without some appreciation of cause and effect, we have no possibility of understanding and controlling the cosmic forces impinging on our lives. Early humans recognized that there are patterns in nature that are predictable—diurnal rhythm, or the movement of the sun; movement of the moon and stars; tides; seasons; animal migration; and plant succession. They were able to exploit these regularities for their own benefit and to avoid potential hazards.

Over time, every human society evolved a culture that inculcated an understanding of its place on Earth and in the cosmos. The collective knowledge, beliefs, languages and songs of each society make up what anthropologists call a worldview. In every worldview, there is an understanding that everything is connected to everything else, that nothing exists in isolation. People have always known that we are deeply embedded in and dependent upon the natural world.

In such a world of interconnectedness, every action has consequences, and since we were part of that world, we had a responsibility to act properly to keep the world in order. Many of our rituals, songs, prayers and ceremonies were reaffirmations of our dependence on nature and our commitment to behave properly. That is how it has been for most of human existence all over the world.

From Naked Ape to Superspecies

But suddenly in the last century, *Homo sapiens* has undergone a radical transformation into a new kind of force that I call a "superspecies." For the first time in the 3.8 billion years that life has existed on Earth, one species—humanity—is altering the biological, physical and chemical features of the planet on a geological scale. That shift to superspecies has occurred with explosive speed through a number of factors. One is population. It took all of human existence to reach 1 billion people in the early nineteenth century. A hundred years later, when I was born, in 1936, there were 2 billion people on Earth. In my lifetime, global population has almost quadrupled. Thus, by virtue of our numbers alone, our species' "ecological footprint" on the planet has grown explosively—we all have to eat, breathe and drink, and clothe and shelter ourselves.

We are now the most numerous mammalian species on the planet, but unlike all the others, our ecological impact has been greatly amplified by technology. Virtually all of modern technology has been developed within the past century, thereby escalating both the scale and scope of our ability to exploit our surroundings. Resource exploitation is fuelled by an exploding consumer demand for products, and the fulfillment of that demand has become a critical component of economic growth. Hyperconsumption in the developed world serves as the model for people in developing countries now that globalization has rendered the entire world population a potential market. Taken together, human numbers, technology, consumption and a globalized economy have made us a new kind of force on the planet.

Throughout our evolutionary past, we were a local, tribal animal. We may have encountered a hundred humans over a range of a few hundred kilometres in a lifetime. We didn't have to worry about tribes on the other side of a mountain or across an ocean; nor did we have to consider the collective impact of our entire species, because our ecological footprint was so much lighter and nature seemed vast and endlessly self-renewing. Our new status of superspecies has been achieved so rapidly that we are only now becoming aware of a new level of collective responsibility, which reflects a dawning realization that taken all together, human activity is the main cause of the current decline in the biosphere's rich diversity and productivity that support all life on Earth.

A Shattered World

As we have shifted our status to superspecies, our ancient under-standing of the exquisite interconnectivity of all life has been shattered. We find it increasingly difficult to recognize the link-ages that once gave us a sense of place and belonging. After all, we are flooded with food and goods that come from all parts of the world, so we scarcely notice that it's the middle of winter when we are buying fresh strawberries and cherries. The constraints of locality and seasons are pushed aside by the global economy. Exacerbating the fragmentation of the world has been the stun-ning shift from predominant habitation in rural village commu-nities to concentration in large cities. In big cities, it becomes easy to assume that we differ from all other species in that we create our own habitat and thereby escape the constraints of nature. It is nature that cleanses water, creates air, decomposes sewage, absorbs garbage, generates electricity and produces food, but in cities, these "ecosystem services" are assumed to be performed by the workings of the economy.

To make matters worse, as we look toward more and more esoteric sources for our information, the context, history and background needed to set new "facts" or events in place are lost, and our world is broken up into disconnected bits and pieces. While we look to science to reveal the secrets of the cosmos, its primary methodology of reductionism focuses on parts of nature. And as the world around us is examined in pieces, the rhythms, patterns and cycles within which those pieces are integrated are lost, and any insights we gain become illusions of understanding and mastery. Finally, as transnational corporations, politics and telecommunications move onto the global stage, the sense of the local is decimated.

This, then, is where we are at the beginning of the third millennium. With explosive speed, we have been transmogrified from a species like most others that live in balance with their surroundings into an unprecedented force. Like a species introduced into a new environment free of constraints, we have expanded beyond the capacity of our surroundings to support us. It is clear from the history of the past two centuries that the path we embarked on after the Industrial Revolution is leading us increasingly into conflict with life-support systems of the natural world. Despite decades of experience in the environmental movement we have not yet turned onto a different path.

The Growth of Environmentalism

Like millions of people around the world, I was galvanized in 1962 by Rachel Carson's eloquent call to action in her book *Silent Spring*. We were swept up in what was to become the "environmental movement." In British Columbia, that meant protesting such threats as the American testing of nuclear weapons at Amchitka in the Aleutian Islands (a protest that gave birth to Greenpeace in Vancouver), clearcut logging throughout the province, proposed offshore drilling for oil, the planned dam at Site C on the Peace River, and air and water pollution from pulp mills. In my mind, the problem was that we were taking too much from the environment and putting too much waste back into it. From that perspective, the solution was to set limits on how much and what could be removed from the biosphere for human use and how much and what could be put back into our surroundings, then make sure to enforce the regulations. So in addition to protesting, marching and blockading, many of us were lobbying politicians to set aside more parks, to enact Clean Water and Clean Air legislation, to pass Endangered Species Acts and to establish the agencies to enforce the regulations. When *Silent Spring* was published in 1962, no government on Earth had a minister or department of the environment.

But Carson's book itself offered evidence of the need for a deeper analysis. As I read the book, I was shocked to realize that the experimental systems scientists study in flasks and growth chambers are artifacts, simplifications meant to mimic reality but lacking the context within which those simplified systems exist and devoid of the rhythms, patterns and cycles that impinge on the Earth. This realization came to me as a profound shock and impelled me to look beyond the lab into the real world.

The more involved I became in environmental issues, the clearer it became to me that my rather simple-minded approach wouldn't work, because we were too ignorant to anticipate the consequences of our activity and to set appropriate limits. Carson's book dealt with DDT. In the 1930s when Paul Mueller, working for the chemical company Geigy in Switzerland, discovered that DDT killed insects, the economic benefits of a chemical pesticide were immediately obvious. Trumpeting the imminent scientific conquest of insect pests and their associated diseases and damage to crops, Geigy patented the discovery and went on to make millions, and Mueller was awarded the Nobel Prize in 1948. But years later, when birdwatchers noted the decline of eagles and hawks, biologists investigated and discovered the hitherto unknown phenomenon of "biomagnification," whereby compounds become concentrated as they are ingested up the food chain. How could limits have been set on DDT in the early 1940s when we didn't even know about biomagnification as a biological process until birds began to disappear?

Similarly, CFCs were hailed as a wonderful creation of chemistry. These complex molecules were chemically inert, so they didn't react with other compounds and thus made excellent fillers in aerosol cans to go along with substances such as deodorants. No one anticipated that because of their stability, CFCs would persist in the environment and drift into the upper atmosphere, where ultraviolet radiation would break off ozone-scavenging chlorine free radicals. Most people had never heard of the ozone layer, and certainly no one could have anticipated the long-term effects of CFCs, so how could the compounds have been regulated? I have absolutely no doubt that genetically modified organisms (GMOs) will also prove to have unexpected negative consequences despite the benefits claimed by biotech companies. But if we don't know enough to anticipate the long-term consequences of human technological innovation, how can its impact be managed? For me as a scientist, this posed a terrible conundrum.

A Way Out

I gained an important insight to free me from this quandary in the late 1970s. As host of the long-running television series *The Nature of Things*, I learned of the battle over clearcut logging on Haida Gwaii, off the coast of British Columbia. For thousands of years, the islands have been home to the Haida people. Forest companies had for years been denuding much of the islands by clearcut logging, which had generated growing opposition. It was a good story, and I proposed to report it. In the early 1980s, I flew to Haida Gwaii to interview loggers, forestry officials, government bureaucrats, environmentalists and Indigenous people. One of the people I interviewed was a young Haida artist named Guujaaw who had led the opposition to logging for years.

Unemployment was very high in the Haida communities, and logging generated desperately needed jobs for the Haida. So I asked Guujaaw why he opposed the logging. He answered, "Our people have determined that Windy Bay and other areas must be left in their natural condition so that we can keep our identity and pass it on to following generations. The forests, those oceans, are what keep us as Haida people today." When I asked him what would happen if the logging continued and the trees were cleared, he answered simply, "If they're logged off, we'll probably end up the same as everyone else, I guess."

It was a simple statement whose implications escaped me at the time. But on reflection, I realized that he had given me a glimpse into a profoundly different way of seeing the world. Guujaaw's statement suggested that for his people, the trees, the birds, the fish, the water and wind are all parts of Haida identity. Haida history and culture and the very meaning of why Haida are on Earth reside in the land.

Ever since that interview, I have been a student learning from encounters with Indigenous Peoples in many parts of the world. From Japan to Australia, Papua New Guinea, Borneo, the Kalahari, the Amazon and the Arctic, Indigenous people have expressed to me that vital need to be connected to the land. They refer to Earth as their Mother, who they say gives birth to us. Moreover, skin enfolds our bodies but does not define our limits because water, gases and heat dissipating from our bodies radiate outward, joining us to the world around us. What I have learned is a perspective that we are an inseparable part of a community of organisms that are our kin.

In 2001, U.S. president Bill Clinton joined with scientists to announce the completion of the Human Genome Project, which elucidated the complete sequence of 3 billion letters in a single human nucleus. While politicians and scientists speculated about the potential benefits of understanding diseases, new drugs and cures for many ailments, the most amazing revelation was all but ignored. Not only is the human genome nearly identical to that of our closest relatives, the Great Apes, as well as our pet dogs and cats, we carry thousands of genes identical to those in fish, birds, insects and plants, a revelation that we share genes with all other life forms to whom we are related by our shared evolutionary history.

Changing Our Perspective

In 1990, my wife, Tara Cullis, and I established an organization that would examine the root causes of ecological destruction so that we could seek alternatives to our current practices. We decided to draft a document that would express the foundation's worldview and perspective and could be offered to the Earth Summit in Rio de Janeiro in 1992. We called it a Declaration of Interdependence. Tara and I formulated a rough draft and asked for input from Guujaaw, ethnobiologist Wade Davis and the children's singer Raffi. When I was working on the first draft, I tried writing, "We are made up of molecules from the air, water and soil," but this sounded like a scientific treatise and failed to convey the simple truth of our relationship with Earth in a powerful, emotional way. After spending days pondering the lines, I suddenly thought, "We *are* the air, we *are* the water, we *are* the earth, we *are* the sun."

With this realization, I also saw that environmentalists like me had been framing the issue improperly. There is no environment "out there" that is separate from us. We can't manage our impact on the environment if we *are* our surroundings. Indigenous people are absolutely correct: we are born of the Earth and constructed from the four sacred elements of earth, air, fire and water. (Hindus list these four and add a fifth element, space.)

Once I had finally understood the truth of these ancient wisdoms, I also realized that we are intimately fused to our surroundings and the notion of separateness or isolation is an illusion. Through reading I came to understand that science reaffirms the profundity of these ancient truths over and over again. Looked at as biological beings, despite our veneer of civilization, we are no more removed from nature than any other creature, even in the midst of a large city. Our animal nature dictates our essential needs: clean air, clean water, clean soil, clean energy. This led me to another insight, that these four "sacred elements" are created, cleansed and renewed by the web of life itself. If there is to be a fifth sacred element, it is biodiversity itself. And whatever we do to these elements, we do directly to ourselves.

As I read further, I discovered the famed psychologist Abraham Maslow, who pointed out that we have a nested series of fundamental needs. At the most basic level, we require the five sacred elements in order to live rich, full lives. But when those basic necessities are met, a new set of needs arises. We are social animals, and the most profound force shaping our humanity is love. And when that vital social requirement is fulfilled, then a new level of spiritual needs arises as an urgent priority. This is how I made the fundamental re-examination of our relationship with Earth that led to *The Sacred Balance*.

The challenge of this millennium is to recognize what we need to live rich, rewarding lives without undermining the very elements that ensure them.

1

Homo sapiens:
Born of the Earth

We need to unearth the old stories that live in a place
and begin to create new ones, for we are storymakers,
not just storytellers. All stories are connected,
new ones woven from the threads of the old.

—ROBIN WALL KIMMERER, *Braiding Sweetgrass*

LIKE ANY OTHER species, human beings have survived because we possess certain traits that have helped us secure a place on Earth. We are not distinguished by an armoury of weapons such as quills, fangs or talons, nor are we possessed of exceptional speed, strength or agility. Our sensory acuity cannot compete with that of other animals; we cannot hear as well as a bat, smell as well as a dog or see as well as an eagle. Yet not only have we survived, we have flourished within a remarkably brief time on the evolutionary scale. The key to our success is the possession of the most complex structure on Earth: the human brain.

Weighing a mere 1.5 kilograms and occupying the space of two fists, the brain consists of 100 billion neurons. Each neuron can form up to ten thousand connections with other cells, thereby creating the potential for more combinations than there are stars in the heavens.

The Brain's Need for Order

Some scientists compare the brain to a relay station that merely coordinates incoming signals and outgoing responses, whereas others see it as an immense computer that processes information and then arrives at an appropriate response. François Jacob, the late French molecular biologist and Nobel laureate, suggested that the human mind is far more; it has a built-in need to *create order* out of the constant flow of information coming from its sensory organs. In other words, the brain creates a narrative, with a beginning, a middle and an end—a temporal sequence that makes sense of events. The brain selects and discards information to be used in the narrative, constructing connections and relationships that create a web of meaning. In this way, a narrative reveals more than just *what* happened; it explains *why*. When the mind selects and orders incoming information into meaning, it is telling itself a story.

But our story is not written at birth; it is fluid throughout our lives. Our story is continually being edited and shaped because our brains *physically* change as we experience our world. Nobel laureate Eric Kandel showed that when animals learn, the actual structure of the neurons and synapses—the relay points that send information from neuron to neuron—change. At birth, for instance, a child has about 2500 synapses per neuron in the cerebral cortex. As the young child grows, he or she is flooded with new experiences, and the brain changes rapidly. After a few years, there are about 15,000 synapses per neuron. As we age, these synapses are "pruned"—weak connections are eliminated, while strong connections are preserved. So all of the connections in our brains are meaningful, part of our history and our experience, each with its own purpose.

This lifelong plasticity of the brain is a relatively new insight. It was thought that beyond a period of rapid change in early infancy, our brains became increasingly fixed as we aged, but it is now clear that the fate of the adult brain is not predestined solely by inheritance. With new brain-imaging technology, we can see how brains work in real time. In pulses of colour, we can see the parts of the brain illuminate when we concentrate, create, dream and experience sensations and emotions. The connections between neurons are strengthened and fine-tuned by experience. Our brains are capable of change throughout our lives; through our experiences, we essentially rewire our brains. The scent of a new box of crayons, the cool squish of mud between our toes, the surge of adrenalin and racing heart when we face fear—through our experiences we craft new metaphors and similes, narrative threads and storylines.

Weaving a Worldview

Early humans identified sequences and repetitions in the world around them—the alternation of day and night, the coming and going of the seasons, the cycle of the tides, the movement of stars across the heavens along predictable paths. People learned to make use of patterns such as the migrations of animals and the seasonal succession of plants; they could "read" the landscape around them to find the things they needed. They investigated the world they lived in and reflected on it with all the built-in ordering capacity of that extraordinary organ, the brain.

If we took the fact of the world's speech seriously—the world speaks!—there would be none of the usual talk about a mechanistic and deterministic science, about a cold, soulless universe, or about an unavoidable conflict between science and the spirit.

Confronting the many voices of nature, we would inquire about their individual qualities and character, we would look for the direction of their expressive striving, and we would struggle to grasp the aesthetic unity of their various utterances—all of which is to say: we would listen for their meanings... The trouble, however, is that we often fail to

pay attention; we never learn the language of the world we inhabit. We
try to master nature while becoming increasingly deaf to her complex
symphony.
  —STEPHEN TALBOTT, Nature Institute senior researcher,
  in "The Language of Nature"

The knowledge of every band of human beings, acquired and
accumulated through generations of observation, experience and
conjecture, was a priceless legacy for survival. All over the world,
small family groups of nomadic hunter-gatherers depended on
skills and knowledge that were profoundly local, embedded in
the flora, fauna, climate and geology of a region. This information was woven together into what anthropologists call a worldview—a story whose subject for each group is the world and
everything in it, a world in which human beings are deeply and
inextricably immersed. Each worldview was tied to a unique
locale and peopled with spirits and gods. At the centre of the
story stood the people who had shaped it to make sense of their
world. Their narrative provided answers to those age-old questions: Who are we? How did we get here? What does it all mean?

Every worldview describes a universe in which everything is
connected with everything else. Stars, clouds, forests, oceans and
human beings are interconnected components of a single system
in which nothing can exist in isolation.

The stars, Earth, stones, life of all kinds, form a whole in relation to each
other and so close is this relationship that we cannot understand a stone
without some understanding of the great sun. No matter what we touch,
an atom or a cell, we cannot explain it without knowledge of the universe.
The laws governing the universe can be made interesting and wonderful
to children, more interesting than things in themselves, and they begin
to ask:

What am I?
What is the task of humanity in this wonderful universe?
—MARIA MONTESSORI, *To Educate the Human Potential*

In such an interdependent universe human beings hold enormous responsibility; each individual is accountable, and every action has repercussions that reverberate far beyond the moment. Past, present and future form a continuum in which each generation inherits a world shaped by the actions of its forebears and holds it in trust for all the generations to come. Many worldviews endow human beings with an even more awesome task: they are the caretakers of the entire system, responsible for keeping the stars on their courses and the living world intact. In this way, many early people who created worldviews constructed a way of life that was truly ecologically sustainable, fulfilling and just.

The Copernican Revolution

For thousands upon thousands of years, people saw the world as a whole and occupied a central place in their worldviews. Then, in 1543, astronomer Nicolaus Copernicus presented a new view of the cosmos, and humanity's place in it, in his monumental book, *On the Revolutions of Celestial Orbs*. His story had the sun at its centre, circled by attendant planets. In 1610, Galileo Galilei published *Sidereus Nuncius*. "The Starry Messenger" announced the discoveries he had made through his "optic tube" about the nature of the moon, the composition of the Milky Way and the "new stars" circling Jupiter—evidence supporting the Copernican hypothesis. Earth was marginalized—reduced to one of many planets in a universe filled with other suns. This revolutionary cosmology demolished the intellectual and moral order of the Western world. As the British poet John Donne lamented:

> The new Philosophy calls all in doubt,
> The Element of fire is quite put out;
> The Sun is lost, and th' Earth, and no mans wit
> Can well direct him where to look for it...
> 'Tis all in peeces, all cohaerence gone;
> All just supply, and all Relation.

The medieval view of the universe as fixed and finite, with humankind enshrined at the centre as God's special creation, was replaced by an infinite universe—a boundless region filled with darkness and space—and human beings were relocated to very inferior quarters: the third planet orbiting a very ordinary star out on a long spiral arm of the astronomically unremarkable Milky Way galaxy. Copernicus pushed us out of the centre, and we've been trying to get back there ever since—claiming power not as part of the web of creation, not even as caretakers, but as masters of a cosmic machine.

We must convince each generation that they are transient passengers on this planet earth. It does not belong to them. They are not free to doom generations yet unborn. They are not at liberty to erase humanity's past nor dim its future.

—BERNARD LOWN and EVJUENI CHAZOV, quoted in P. Crean
and P. Kome, eds., *Peace, a Dream Unfolding*

Two centuries after Copernicus, the great physicist Isaac Newton discovered laws governing the movement of bodies and the behaviour of light that appeared to apply everywhere in the universe. The cosmos, he concluded, is like an immense clock, a complex mechanism whose basic components and principles could be revealed and examined through science. According to this view, nature is a machine and is no more than the sum of its parts; scientists could add fragments of information together like the pieces of a jigsaw puzzle until they obtained a comprehensive picture of the whole. Thus, to those accepting Newton's ideas, the natural world, like any other machine, is knowable, adjustable, manageable. And above all, it belongs to the people who control it.

Charles Darwin's magnificent opus, *On the Origin of Species,* was the biological equivalent of the Copernican Revolution. By replacing the moment of divine creation, when God made Adam and Eve in his own image and gave them dominion over the

The Sun Is My Father, the Earth Is My Mother

For the Desana people of the northwestern Amazon, *page abe,* the Sun, was the creator of the universe. Moon was his twin brother. One of their origin stories describes Sun plunging his long rattle deep into the Earth until it penetrated the fertile paradise of the lower world, called *ahpikondia,* the River of Milk. Then, holding the stick vertically so that it cast no shadow, *page abe* released droplets of supernatural sperm, which cascaded down it, inseminating the Earth and creating men. Leaving the cosmic womb by climbing up the Sun's rattle, they emerged on Earth's surface as fully formed Desana men. This act of creation gave shape to the world, as the power of the Sun's yellow light gives it life and stability...

According to the Desana, this inherent stability of the natural world is rooted in a vast web of reciprocal relationships that have always existed between all elements of nature. A reciprocity between the Earth, with its mountains and forests and rivers, and the first forms of life—the animals, the plants, the Desana people—exists in harmony with all of the rest of the universe. As traditional Desana tell it, the Sun planned his creation well, and it was perfect. ☐

entire Earth, with a long-running family saga that includes apes and chimpanzees, Darwin shoved the human species off its pedestal. Succeeding generations of evolutionary biologists denied humanity its final claim to eminence by showing that natural selection does not necessarily lead to increasing levels of complexity and greater intelligence. As Stephen Jay Gould argued so well in *Wonderful Life*, the kind of species that evolves at any moment depends on circumstances and is not part of a progression that follows an underlying principle. There is no splendid evolutionary ladder leading steadily onward and upward toward *Homo sapiens*.

Descartes's famous phrase *Cogito, ergo sum* ("I think, therefore I am") encapsulates the belief that self-consciousness, or awareness, is the great achievement of the human species, a property that is unique to humankind and elevates us above all other life. Even that conceit crumbles as neurobiologists explore the electrochemistry, physiology and anatomy of neurological systems. According to Donald R. Griffin:

> Human consciousness and subjective feelings are so obviously important and useful to us that it seems unlikely that they are unique to a single species. This assumption of a human monopoly on conscious thinking becomes more and more difficult to defend as we learn about the ingenuity of animals in coping with problems in their normal lives.

Human beings, especially in Western societies, have long nurtured a sense of superiority over other animals. We were said to have been created in the image of God. We are thought to be the only creatures that have self-awareness, that make tools, that reason in the abstract, that have empathy for others. Yet over and over, as scientists begin to observe and interpret animal behaviour, the underpinnings of our alleged superiority have been pulled away.

A Candid Conversation between Two Species

The Man: *I am the predilect object of Creation, the centre of all that exists...*

The Tapeworm: *You are exalting yourself a little. If you consider yourself the lord of Creation, what can I be, who feed upon you and am ruler in your entrails?*

The Man: *You lack reason and an immortal soul.*

The Tapeworm: *And since it is an established fact that the concentration and complexity of the nervous system appear in the animal scale as an uninterrupted series of graduations, where are we cut off? How many neurons must be possessed in order to have a soul and a little rationality?*

—SANTIAGO RAMÓN Y CAJAL, *Recollections of My Life*

Darwinian evolution has cast us as the children of chance, creatures with enough self-awareness and wit to recognize ourselves as a kind of cosmic joke. From Copernicus to Darwin to the reflections of modern eminent scientists, in the Western world *Homo sapiens* has undergone a relentless diminuendo, ending up as just another species that happened to evolve way out in the heavenly boonies.

Severing the Connections

Whereas traditional worldviews see the universe as a whole, science produces information that can never, almost by definition, be complete. Scientists focus on *parts* of nature, attempting to isolate each fragment and control the factors impinging on it. The observations and measurements they make provide a profound understanding of that bit of nature. But what is ultimately acquired is a fractured mosaic of disconnected bits and pieces, whose parts will never add up to a coherent narrative.

The very act of focusing on and isolating a part of nature removes the context within which that fragment had meaning or significance. We lose sight of the rhythms, patterns and cycles within which that piece exists. And so we create an artifact that informs us very little about the properties and behaviours of that bit of nature in the real world.

Furthermore, the Newtonian method of understanding a whole system by adding together the properties of its parts has turned out to be fundamentally flawed. Over time it has become clear that at every level the effort to know the whole from the parts is doomed. At the most elementary level of matter, physicists examining parts of atoms early in the twentieth century created a solar system–like model with discrete protons and neutrons at the centre analogous to the sun, and electrons orbiting the nucleus like the planets. Quantum mechanics destroyed this comforting model by replacing it with an atomic image whose components could only be predicted statistically. That is, the position of a particle could not be defined with absolute certainty but only by statistical probability. If there is no absolute certainty at the most elementary level, then the notion that the entire universe is understandable and predictable from its components becomes absurd.

Far worse, different parts of the real world interact synergistically when placed together. As Nobel laureate Roger Sperry points out, new properties that arise from complexes cannot be predicted from the known properties of their individual parts. These "emergent properties" only exist within the whole. So we can never learn how whole systems work simply by analyzing each of their components in isolation. One cannot, for instance, examine a single neuron and imagine the complexity of the human brain, or study a single molecule of H_2O and envision the wetness of water. In a reductionist view, cause-and-effect is linear: a change at one end results in a corresponding change at the other. But most systems are not linear—they are complex and interwoven, dynamic, synergistic and interdependent. Variable is layered upon variable, and in the world outside of a lab, the repercussions from all of the possible interactions seem infinite.

The field of complexity science looks at systems—be they meteorological, biological, physical, even economic or cultural—and the inherent, and usually unpredictable, structure that emerges from them. How do the parts give rise to collective behaviours? How do systems interact with their environments? At some point, perhaps at high density, a chaotic system of individuals undergoes a transition to order. And with this order, the complex system is highly adaptive, with a heightened capacity to respond to a constantly changing and unpredictable world.

Stuart Kauffman, a pioneer in the field of complexity science, contends that self-organization of complex systems is one of the great principles of nature, alongside the Darwinian drivers of chance and selection. He explains:

> ... natural selection is important, but it has not labored alone to craft the fine architectures of the biosphere, from cell to organism to ecosystem. Another source—self-organization—is the root source of order. The order of the biological world... is not merely tinkered, but arises naturally and spontaneously because of these principles of self-organization...

At a certain point, self-organization or unexpected qualities emerge—a collection of molecules into a cell, a collection of sounds into speech, a collection of individual ants into a colony. In essence, an unexpected whole emerges from simple parts.

Revealing How Little We Know

It could be argued that even if statistical uncertainty and synergism prevent a Newtonian exposition of the universe, scientists can nevertheless continue to search for universal principles that apply to different levels of life—subatomic, atomic, molecular, cellular and so on. The problem is that despite the impressive scientific gains made in this century, what we know is utterly minuscule compared with everything that remains unknown or not understood.

Identifying a species merely indicates that the taxonomic position of a specimen has been tracked down so that it can be given a name. It does not mean that anything is known about its numbers, distribution, basic biology or interaction with other species—studies that might take a human lifetime to complete for each species.

Estimating the number of species on Earth is challenging at best and strongly biased to "charismatic megafauna" (large creatures, such as tigers, bears or whales, that people find most engaging), temperate ecosystems and organisms directly relevant

The Truffle-Eucalyptus-Poteroo Connection

Environmentalist Ian Lowe of Griffith University in Australia relates a story that illustrates the exquisite and unpredictable interconnectedness of life's components. In a study of truffles that grow in the dry eucalyptus forest of New South Wales, it was found that the truffles perform a service for the trees near which they are found. Because both truffles and trees extract water and minerals from the soil, trees with truffles in their roots obtain more water and minerals and grow better than those without. The truffles are a favourite food of the long-footed poteroo, a marsupial that is now classified as rare, which then excretes the spores of the truffles and thereby enhances the health of the forest. Poteroo, truffle, eucalypt—three very different species of mammal, fungus and plant—are all bound together in a remarkable web of interdependence. □

(or at least intriguing) to humans. The *Millennium Ecosystem Assessment* estimates the total species on Earth as somewhere between 5 and 30 million. Of these species, fewer than 2 million have been described scientifically.

Just one fistful of moss from the forest floor, for example, can be home to 150,000 protozoa, 132,000 tardigrades, 3000 springtails, 800 rotifers, 400 mites, 200 larvae and 50 nematodes. In the old-growth forests of the upper Carmanah Valley on Vancouver Island, biologist Neville Winchester collected 1.4 million specimens, which included ten to fifteen thousand species of invertebrates (mostly insects), about one-third of the known species in all of Canada. Of these, at least five hundred were new to science.

The sheer numbers of nematodes alone could inspire nightmares and B movies. These microscopic roundworms—most of which are parasitic but some of which are free-living species in soil and water—are so numerous that it is estimated that four out of every five animals on Earth are nematodes. Consider this scene imagined by nematologist Nathan Cobb:

> If all the matter in the universe except nematodes were swept away, our world would still be recognizable... we should find its mountains, hills, vales, rivers, lakes, and oceans represented by a film of nematodes. The location of towns would be decipherable, since for every massing of human beings there would be a corresponding massing of certain nematodes.

Just as the soil beneath our feet is poorly understood, so too is life in the ocean. The ten-year Census of Marine Life strives to change this. The census, which includes 1700 experts from 73 nations, was established to assess the diversity of marine life worldwide, as well as how species interact with each other and their ecosystems. Completed in 2010, the census adds to our understanding of the ocean's diversity while also indicating the magnitude of our ignorance. The first-ever census of the Gulf of Maine, for instance, yielded over 50 per cent more species than estimated, and in a survey of the South Atlantic, nearly 30 per cent of the species sampled were new to science. A thirty-day expedition to the Arctic Ocean's Canada Basin yielded twelve new species.

Figures such as these show us just how little power we really possess. We are a long way from being able to make even an educated guess as to how to manage natural systems, especially ones as complex as forests, wetlands, prairies, oceans or the atmosphere. The Nobel laureate Richard Feynman once observed that trying to understand nature through science is like trying to figure out the rules of chess as you watch a game being played—but you can only see two squares at a time.

Similarly, our knowledge of the geological and geophysical structure of the planet is minuscule and fragmentary. Scientists are sometimes criticized because they can't make up their minds about the rate and intensity of the global warming currently under way. Meteorologists have difficulty predicting local weather from day to day, so the problems in predicting climate across decades should come as no surprise. Our knowledge base is so primitive that merely tweaking assumptions here or there in computer models of climate change can alter predictions from a 1.5° to 6.0°C rise by this century's end. That is not an indictment of scientists but indicates that there are gaps in our knowledge large enough for the future of the planet to fall through.

Another problem with the Newtonian method and science as a whole is that scientists seek principles that are universal and replicable anywhere and any time—thereby severing them from temporal and geographic specificity. The technological prowess developed in biotechnology to manipulate DNA is most impressive. Plants have been developed in labs to resist pests or chemical pesticides. But labs or growth chambers, where variables such as temperature, humidity, nutrition and other species are rigorously controlled, do not mimic or anticipate the kinds of conditions encountered in India, Africa or specific locales in the United States. By attempting to observe fragments of nature objectively and without emotion, scientists extirpate the passion and love that piqued their curiosity in the first place, often to discover that they have so objectified the focus of their attention that they no longer care.

Albert Einstein was asked one day by a friend "Do you believe that absolutely everything can be expressed scientifically?" "Yes, it would be possible," he replied, "but it would make no sense. It would be description without meaning—as if you described a Beethoven symphony as a variation in wave pressure."

—RONALD W. CLARK, *Einstein: The Life and Times*

Scientism, the aura of authority carried by scientists, has made us believe that knowledge obtained by scientists is the ultimate authority, that as we accumulate information, our capacity to understand, control and manage our surroundings will grow correspondingly. But the basic principle of scientific exploration contradicts this faith: knowledge comes from empirical observations, which are "made sense of" by hypotheses, which in turn can be experimentally tested. All information is open to being disproved. As Jonathan Marks has pointed out:

... the vast majority of ideas that most scientists have ever had have been wrong. They have been refuted; they have been disposed of. Further, at any point in time, most ideas proposed

by most scientists will ultimately be refuted and disposed of...
Science, in other words, undermines scientism.

Unfortunately, we seem to be on a journey that gets longer
with every step we take. Science is strong on *description;* we know
so little that scientists make discoveries everywhere they look.
But each discovery merely reveals the magnitude of our ignorance; far from filling in the picture, these discoveries show us
just how much still remains to be learned.

The total knowledge base currently accumulated by scien-
tists is still so limited that it can rarely be *prescriptive;* it is almost

Reading the Book of Life

With our advances in technology—sieves so fine they can separate organisms from between grains of sand, electron microscopes
that can image cross-sections mere microns thick and computers
that can process data at unfathomable speeds—the adage that we
know less and less about more and more becomes ever more true.
As each new technological advance reveals incredible details about
the minutiae of our world, a flood of further questions arises. As
an example, take the Human Genome Project (HGP), which set
out to decipher and sequence the entire 3 billion letters of the
genetic code within human DNA.

At its inception in 1990, researchers thought they would find
about a hundred thousand human genes. In the end, they identified twenty to twenty-five thousand. Given that the number of
genes was considered a measure of genetic complexity, this relatively low number came as a shock to many. As a comparison,
the nematode *C. elegans* has about twenty thousand genes, and,
what's more, about 40 per cent of human genes are virtually

impossible to generate scientifically based policies or solutions for managing our surroundings when we know so little. It is as if we are standing in a cave holding a candle; the flame barely penetrates the darkness, and we have no idea where the cave walls are, let alone how many more caves there are beyond. Standing in the dark, cut off from time, and place, and from the rest of the universe, we struggle to understand what we are doing here alone.

To become human, one must make room in oneself for the wonders of the universe.

—SOUTH AMERICAN INDIGENOUS SAYING

identical in their basic structure to those of this "simple" creature. The recent completion of the chimpanzee genome shows that we share about 98 per cent of our DNA sequences with our closest living relative. But what do these figures really tell us?

Clearly, defining the code, essentially finding the letters of our genetic blueprint, is just the beginning. Eric Lander, a lead scientist with the HGP, says that the achievement of sequencing the human genome is like having evolution's notes about life for the last 3 billion years. The problem is, we can't read the notes. As Lander explains, "It's as if we got the keys to this amazing library and we're going and we're pulling down all these great volumes... but, of course, we're kindergarteners reading this stuff."

With the completion of the HGP, we have a list of the genes that make human beings (and several other organisms), but we have no idea how to assemble an organism out of these parts. Decoding the human genome provides us with one piece of a very complex puzzle as we continue on the journey to discover what it means to be human. ☐

The scientific method has reduced the world into disciplines, and within each discipline the focus of study is further narrowed and fragmented. This objective, precise scientific methodology has gained us detailed knowledge about snippets of our world, but this system also has its limits and unimagined consequences. When our world is reduced to minutiae, it is easier to treat everything as a commodity. When we reduce an organism to a list of genes, for instance, the context and wholeness within and around that organism is gone, and the next step could be (and has been) the movement of one gene to an entirely unrelated organism. The results of such experiments are highly unpredictable because we are dealing with living organisms, each with its own complex genome.

Biologist Brian Goodwin suggested that science would benefit from acknowledging a way of knowing about the world that includes intuition. By trying to remove the subjectivity of humans from science, we have created a construct that is not realistic; we have tried to remove the scientist and his or her emotions and intuition from the equation.

Goodwin's antidote is a shift toward more holistic science—to study wholes and their relation to their components rather than to reduce the world into small, seemingly unrelated parts. He explained:

> [The intuitive way of knowing is] not something that's vaguely subjective and artistic, it's a definite way of knowing the world. In fact, it's absolutely essential to creative science. All the great scientists, Einstein, Feynman, you name them, would say intuition is the way they arrived at their basic insights, their new ways of putting parts together into coherent wholes. The famous guys are allowed to say this. The rest of us have to pretend we're really basing everything on hard fact, proceeding to generalize by induction... not seeing a new whole intuitively.

When scientific endeavour is severed from its historical and local context, it becomes an activity carried out in a void—a story that has lost its meaning, its purpose and its ability to touch and inform.

Consumption to Satisfy Our Needs

Losing our place in the scheme of things, our specialness, even our gods, has left us with a great ache, a loss, a loneliness, a terrible emptiness. One way we have attempted to fill the void is with a new sacrament: the ritual exchange of money for goods in the temples of the marketplace. As physicist and author Brian Swimme puts it:

> Humans gather together and learn the meaning of the universe, our cosmology. Now, we gather together and watch TV ads. Every ad is a cosmological sermon—the universe is a collection of objects to be fashioned into items for our consumption, and the role of humans is to work and buy objects.

Like poor savage Caliban, teased by scraps of half-heard music, the "sounds and sweet airs, that give delight and hurt not," that drift about his island prison, we mistake hints and memories of an ancient, long-lost harmony for some more immediate need; enslaved like Caliban to the material world, we dream of the healing power of riches, and when we wake, we "[cry] to dream again."

The rise in our collective and individual demand for consumer goods began in earnest in the twentieth century. As early as 1907, economist Simon Nelson Patten espoused an idea that was to consume the modern world: "The new morality does not consist in saving but in expanding consumption." As Paul Wachtel put it:

> Having more and newer things each year has become not just something we want but something we need. The idea of more, of

ever increasing wealth, has become the center of our identity and our security, and we are caught up by it as the addict by his drugs.

The Great Depression of the 1930s came to an end because the Second World War provided a massive economic jolt. American industrial might burned at white heat to support the war effort, but as victory loomed, the business community worried about how to keep the economy going. The answer was consumption. Shortly after the Second World War, retailing analyst Victor Lebow declared:

> Our enormously productive economy... demands that we make consumption our way of life, that we convert the buying and use of goods into rituals, that we seek our spiritual satisfaction, our ego satisfaction, in consumption... We need things consumed, burned up, worn out, replaced, and discarded at an ever-increasing rate.

In 1953 the chairman of President Eisenhower's Council of Economic Advisors stated that the American economy's "ultimate purpose" was "to produce more consumer goods." This strategy was successful, and the shift from buying what we need to buying what we want began in earnest and escalated to the point at which we find ourselves today: with megamalls, big-box stores and bumper stickers declaring that we were "Born to Shop." Consumerism is so entrenched as the driver of Western economies that shopping can even be couched as one's patriotic duty. Within weeks of the tragedies of September 11, 2001, it was clear that consumer confidence was shaky and economies were lagging. President George Bush, as well as other Western leaders, began to encourage spending as a meaningful way for people to help. Canada's prime minister, Jean Chrétien, reminded us that interest rates were low "so it is time to go out and get a mortgage, to buy a home, to buy a car," and President Bush encouraged people to help restore confidence in the airline industry: "Get on board. Do your business around the country. Fly and enjoy America's great destination spots. Get down to Disney World in Florida." Although these actions would, of course, stimulate businesses, factories and jobs, it is a sad statement about Western society when leaders suggest that the most meaningful way to help at a time of such tragedy is to shop till you drop.

When products are made to last, businesses eventually run out of customers. Planned obsolescence is one solution; another approach is to bring forth new lines of products carrying an array of bells and whistles, a strategy exemplified by the automobile, computer and fashion industries. Constantly redefining potential markets is yet another strategy, expanding to the developing world or specific ethnic groups, for example. Coca-Cola president Donald R. Keough expressed a quasi-religious attitude toward market opportunity: "When I think of Indonesia—a country on the Equator with 180 million people, a median age of 18, and with a Moslem ban on alcohol—I feel I know what heaven looks like."

All demographic groups—elders, "yuppies," "baby boomers," for example—are prime targets for marketers, but in a relatively recent shift, children have been clearly placed in the bull's eye. Between 1983 and 1997, the amount of money spent marketing to American children increased from $100 million to $12 billion. By 2005, the total closed in on $15 billion. And all of this marketing seems to be paying off. In 1989 children ages four to twelve spent $6.1 billion in consumer purchases. In 2002, this figure was $30 billion, an increase of 400 per cent.

Even before children have turned on the television or stepped outside the front door, it's likely they've been subjected to advertising. Today, pyjamas, cereal, bedsheets, curtains, wallpaper and even toothpaste, bandages, soap and shampoo advertise images from favourite television or movie characters. In many cities, the advertising bombardment doesn't stop when the children reach the schoolyard. Dwindling budgets for education have also introduced advertising into the classroom. In the United States, almost 7 million students (and nearly 30 per cent of teenagers) watch Channel One—a daily news broadcast with two minutes of advertisements—in their classrooms every morning in exchange for the use of "free" equipment such as televisions, video monitors, computers and other electronics. The advertisers know a captive and lucrative audience when they see one and so are willing to pay twice as much for an ad spot on Channel One as they would on a regular network outside of the classroom.

As Juliet Schor, author of *Born to Buy*, says: "We have become a nation that places a lower priority on teaching its children how to thrive socially, intellectually, even spiritually, than it does on training them to consume. The long-term consequences of this development are ominous."

The English language is made up primarily of nouns, somehow appropriate for a culture so obsessed with things... English encodes human exceptionalism, which privileges the needs and wants of humans above all others and understands us as detached from the commonwealth of life.
—ROBIN WALL KIMMERER, "Speaking of Nature"

In order to feed the ever-growing consumer demand, constant economic growth is required. The rationale for growth in consumption and the economy is summarized by P.M. McCann and his co-authors: "Growth leads to increasing wealth and this, through the market system, provides the basis for the satisfaction of all human needs." It is an astonishing assertion—that wealth can satisfy all human needs. This is a far cry from the lessons our grandparents taught about the virtues and the pleasures of thrift, about the true values of life and about the sources of happiness.

The purchase of a new product, especially a "big ticket" item such as a car or computer, typically produces an immediate surge of pleasure and achievement, and often confers status and recognition upon the owner. Yet as the novelty wears off, the emptiness threatens to return. The standard consumer solution is to focus on the next promising purchase.

—ALLEN D. KANNER and MARY E. GOMES, "The All-Consuming Self"

Are we better off as a society, now that we've become professional consumers, driving economic growth endlessly ahead of us? It depends on what is considered "better." The United States best exemplifies the fully fledged consumer society; how do its citizens measure up to their national ideals? The nation that values youth and thinness is the most obese in the world. The place where the dollar rules has more disparity between rich and poor than any other industrialized nation. Although peace is one of its highest ideals, the United States is well known for violence. More people use drugs regularly in this land of opportunity than in the rest of the world put together. And more people per capita are imprisoned in the land of the free than in any other Western country. Longer working hours, higher levels of stress, failing families, drug addiction, children at risk—these may be to some extent the pathology of consumerism. Let loose in the world's biggest store, people suffer from various ills: the plague of having too much, the rage and jealousy of those who cannot buy the merchandise.

Nevertheless, governments of all countries continue to hold up economic growth, on which consumerism depends, as the key to their well-being. Countries such as India and China are determined to achieve our level of affluence—a sixteen- to twentyfold increase in their current consumption. Imagine those populations with the same per capita car ownership as the United States; the ecological consequences would be catastrophic. Yet why should they set their sights lower than ours? In April 1990 José Lutzenberger, Brazil's environment minister at the time, addressed an international meeting of world parliamentarians in Washington, D.C. He pointed out that if ownership of private cars throughout the world equalled that of the United States or Japan, cars would reach a total of 7 billion partway through the twenty-first century. He exclaimed, "But this is unthinkable!" When increasing consumption becomes part of our definition of "progress," and ownership of objects is the chief path to happiness, no nation can ever call a halt to economic growth.

Many have recognized that increased consumption is not the key to happiness or satisfaction. At the beginning of the American experiment in democracy, one of the authors of the Constitution of the United States, Benjamin Franklin, said: "Money never made a man happy yet, nor will it. There is nothing in its nature to produce happiness. The more a man has, the more he wants. Instead of filling a vacuum, it makes one."

In 1994, when my father was eighty-five, I moved into his home to care for him in the final weeks of his life. He was dying of a form of cancer that was sapping his strength but caused little pain. Dad was lucid, aware of his impending death and unafraid. As my family and siblings came to spend time with him in the last days, we regaled each other with stories—experiences with family, friends and neighbours that enriched our lives. There was never discussion of possessions—money, big cars or houses, closets full of clothes—only people and shared adventures, which is what life really boils down to.

Consumer World

· A Lancet study conducted in New Zealand between June 2014 and June 2015 found children were exposed to an average of 554 brands per ten-hour day, nearly a brand a minute, mostly in schools and at home.

· At least one-third of all food produced is wasted each year—about 1.3 billion tonnes—while 1 billion people remain under-nourished and another 1 billion stay hungry.

· Every year, 70 million trees in endangered and ancient forests are cut down and replaced by tree plantations in order to make wood-based fabrics like rayon, modal and viscose.

· At the very time that family sizes have dropped precipitously in North America, the average house size has more than doubled from about 100 square metres in 1949 to 210 square metres in 2020.

· From 2018 to 2021, the use of smartphones to shop online more than doubled—in part because of the covid-19 pandemic, but the trend is expected to continue.

· suv emissions nearly tripled in the 2010s, owing to their increasing global popularity. Today, emissions from suvs are comparable to those of the entire international shipping industry.

· From 2001 to 2020, according to Global Forest Watch, global tree cover decreased by around 10 per cent, mostly because of development and agriculture. □

Cutting the Nature Connection

At the same time that consumption and economic growth have been steadily increasing, more and more people have moved from the countryside to the city. Over half of all people throughout the world now live in cities, and the largest influx to cities is occurring in the developing countries.

The most destructive aspect of cities is the profound schism created between human beings and nature. In a human-made environment, surrounded by animals and plants of our choice, we feel ourselves to have escaped the limits of nature. Weather and climate impinge on our lives with far less immediacy. Food is often highly processed and comes in packages, revealing little of its origins in the soil or telltale signs of blemishes, blood, feathers or scales. We forget the source of our water and energy, the destination of our garbage and our sewage. We forget that as biological beings we are as dependent on clean air and water, uncontaminated soil and biodiversity as any other creature. Cut off from the sources of our food and water and the consequences of our way of life, we imagine a world under our control and will risk or sacrifice almost anything to make sure our way of life continues. As cities continue to increase around the world, policy decisions will more and more reflect the illusory bubble we have come to believe is reality.

Just a few generations ago, it was the norm for children to head out to play in the morning or after school, coming home only when hungry or hurt. They might have mucked about in a nearby pond or climbed in a tree fort or kicked balls in a field, but most children were outdoors, and chances are, there was a wild area—even a vacant, overgrown urban lot—nearby. Today, many children have lost the connection with the outdoors. "Playtime" is often highly structured or involves parent-supervised activities, and parents are less willing to send their children out without knowing where they are and what they're doing. (Even playgrounds are being eliminated, largely because of the increasingly litigious nature of North Americans.) As Richard Louv writes in *Last Child in the Woods,* there is a bogeyman around every corner, be it traffic, strangers, crime or even nature itself.

Increasing urbanization has diverted streams into culverts, paved over the swamps and plunked housing developments where forests once grew. Increasingly, nature is imagined to be "out there," perhaps in parks or small green spaces, but not in the ditch down the street, and usually it entails getting into a car to get there. The current generation of children are some of the most disconnected from nature ever in our history. They know how to use a computer mouse but might never have seen a wild one. Children might have heard of climate change and endangered species but probably can't name even a few native plants from their home. As Louv says, "A kid today can likely tell you about the Amazon rainforest—but not about the last time he or she explored the woods in solitude, or lay in a field listening to the wind and watching the clouds move."

Perhaps such thoughts can be brushed off as nostalgic or sentimental, but when direct experience with nature is limited, so is our emotional connection to the places that ultimately sustain us.

As we distance ourselves further from the natural world, we are increasingly surrounded by and dependent on our own inventions. We become enslaved by the constant demands of technology created to serve us. Consider our response to the insistence of a ringing telephone or our behavioural conformity to the commands of computers. Divorced from the sources of our own existence, from the skills of survival and from the realities of those who still live in rural areas, we have become dulled, impervious, slow.

Through our loss of a worldview, our devotion to consumerism and our move into the cities and away from nature, we have lost our connection to the rest of the living planet. We must find a new story, a narrative that includes us in the continuum of Earth's time and space, reminding us of the destiny we share with all the planet's life, restoring purpose and meaning to human existence.

If we had a keen vision and feeling of all ordinary human life, it would be like hearing the grass grow and the squirrel's heart beat, and we should die of that roar which lies on the other side of silence. As it is, the quickest of us walk about well wadded with stupidity.

—GEORGE ELIOT, *Middlemarch*

Finding a New Story

If modern science has not created a coherent worldview, and consumerism does not fill the emptiness of life lived without one, how can we restore our connection to the rest of life on Earth and live rich, fulfilling lives? Where can we find a new story?

A human being is part of the whole, called by us the universe. A part limited in time and space. He experiences himself, his thoughts and feelings, as something separate from the rest, a kind of optical delusion of his consciousness. This delusion is a kind of prison for us, restricting us to our personal desires and to affection for a few persons nearest to us. Our task must be to free ourselves from this prison by widening our circle of compassion to embrace all living creatures.

—ALBERT EINSTEIN, quoted in P. Crean and P. Kome, eds.,
Peace, a Dream Unfolding

We have much to learn from the vast repositories of knowledge that still exist in traditional societies. This was suggested in a report in 1987 by the World Commission on Environment and

Development headed by Norwegian prime minister Gro Harlem Brundtland. Entitled *Our Common Future,* it acknowledged the inability of scientists to provide direction in managing natural resources and called for recognition of and greater respect for the wisdom inherent in traditional societies:

> Their very survival has depended upon their ecological aware-ness and adaptation... These communities are the repositories of vast accumulations of traditional knowledge and experience that links humanity with its ancient origins. Their disappear-ance is a loss for the larger society, which could learn a great deal from their traditional skills in sustainably managing very com-plex ecological systems. It is a terrible irony that as formal devel-opment reaches more deeply into rainforests, deserts, and other isolated environments, it tends to destroy the only cultures that have proved able to thrive in these environments.

In this new millennium, after a century of explosive growth in science and technology, it is fitting that leading members of the scientific community are starting to understand that science alone cannot fulfill humankind's needs, and can often be destruc-tive. One of the most positive developments in science is the growing recognition that Indigenous knowledge is also needed to deepen our understanding of the world and our place in it. In many ways, Indigenous knowledge is more encompassing and profound than science. It embraces gratitude, love, respect and a deep sense of responsibility. Scientists scrub such emotions in the name of objectivity and try to find universal principles that can be replicated anywhere. Indigenous knowledge is the accu-mulated observations, trials and errors, successes and failures of a people living in place. It's critical for a people's survival and has been tested by their presence over thousands of years.

It's also deeply subjective and meaningful, considering the web of relationships with all other animal and plant species—and air, water, soil and sunlight. The element of reciprocity, of responsibility, is missing in contemporary science, and society in general.

Science can be a way of forming intimacy and respect with other species that is rivaled only by the observations of traditional knowledge holders. It can be a path to kinship.

—ROBIN WALL KIMMERER, *Braiding Sweetgrass*

Once, our worldview embedded each of us within a world in which all the parts were intricately interconnected. Each of us could be at the centre of this multidimensional web of inter-connections, "trapped," in a sense, by our total dependence on all of the strands enfolding and infusing us, yet deriving the ulti-mate security of place and belonging. But the inventiveness of our extraordinary brain has freed us from the constant need to make a living from our immediate surroundings. We have ampli-fied our mental reach by science, engineering and technology, and by means of computers and telecommunications we have developed an unprecedented capacity for collecting and assess-ing information. The challenge now is to use these techniques to rediscover our connections to time and space, our place in the biosphere. Scientists know as much as anyone about the won-der, mystery and awe that surround and inhabit us. With their help we can search for a new understanding of the world, regain a sense of its fecundity, its generosity and its welcome for the errant species we've become.

As scientists, many of us have had profound experiences of awe and rev-erence before the universe. We understand that what is regarded as sacred is more likely to be treated with care and respect. Our planetary home should be so regarded. Efforts to safeguard and cherish the environment need to be infused with a vision of the sacred.

—CARL SAGAN et al., "Preserving and Cherishing the Earth:
An Appeal for Joint Commitment in Science and Religion"

Can we combine the descriptive knowledge of modern science with Indigenous knowledge and ancient wisdom to create a new worldview, a story that includes us all? We may find some clues by looking over our shoulder at ideas from our own past. Philosophers from ancient Greece believed the material universe was divisible into just four elements—air, water, earth and fire. In these elements opposite qualities were combined—heat and cold, wet and dry, heavy and light—intermingled in infinitely varied proportion, moving and changing, perpetually at war with each other, yet capable of proper balance. That dynamic balance formed the structure, infused the life of all creation, at every level. Each human being was compounded of those four elements in varying proportions—air, water, earth and fire interacting to generate and sustain life. These ideas persisted for more than two thousand years, defining Shakespeare's intellectual world, and that of writers and thinkers generations after him. Today in altered form they seem newly relevant. Air, water, earth and fire—these are the substances that support all life. Together with the sum total of that life, they maintain the planet, keep it fit for life. As we explore each element in turn, looking at its origins, its function on the planet and our intimate relationship with it, we will begin to understand our indissoluble connection to the centre. We are creatures of the Earth, and everything we learn about the Earth teaches us about ourselves.

... *there is a continuous communication not only between living things and their environment, but among all things living in that environment. An intricate web of interaction connects all life into one vast, self-maintaining system. Each part is related to every other part and we are all part of the whole, part of Supernature.*

—LYALL WATSON, *Supernature*

2

The Breath
of All Green Things

Our next breaths, yours and mine, will sample
the snorts, sighs, bellows, shrieks, cheers and spoken prayers
of the prehistoric and historic past.

—HARLOW SHAPLEY, *Beyond the Observatory*

AN INVISIBLE FORCE surrounds us, fills us up and gives us
life. We know it by many names: air, breath, spirit, wind,
atmosphere, the sky, the heavens. Sometimes we even call it God;
in myth and poetry we assign divine powers to the air. "Wild air,
world-mothering air," Gerard Manley Hopkins calls it, the "wild
web, wondrous robe" that wraps the planet like God's mercy. Air
is the creative force, the spirit moving over the face of the waters
in the book of Genesis, the Word of God breathing life into the
world according to Psalm 33, the divine command that sets so
many other creation stories into motion. It is the first of the elements that create and support our lives.

Air also embodies ideas in speech and language, in song and
in the sweet airs of music. In English, as in other languages, a web
of words celebrates the sacred status of air. Look at how the word
"spirit" expands from its Latin source, *spiritus*, meaning "breath,"
"air," into so many other lively meanings—the soul, the animating

principle, intelligence, emotional vigour, liveliness, essence or distilled extract—each one in opposition to deadness or dullness. From the same root comes "inspiration," which gives birth to a new idea, and "expiration," which signals the end of life. Our language knows better than we realize the vital nature of the air we breathe. It is the whirlwind and the breeze, a moving ocean of invisible forces in which we swim all the days of our lives, from our first gasp at birth to our last, slow exhalation at death.

Air is our element; we live inside the atmosphere, the envelope of mixed gases that form the outer layer of the planet. Two thousand years ago Plato pointed out that we are "dwelling in a hollow of the earth, and fancy that we are on the surface... But the fact is, that owing to our feebleness and sluggishness we are prevented from reaching the surface of the air." Since then technology has partially overcome our sluggishness, but when we leave our element we have to take some with us. Oxygen tanks will keep us breathing in the thin air of mountain peaks and under water; in space astronauts are protected from instant death by the bubble of air in their capsule or suit.

At first you are awed by the splendour, by the beauty, of the planet and then you look down and you realize that this one planet is the only thing we have. Every time the sun comes up and goes down... and for us that's sixteen times a day... you see a thin, thin, thin layer just above the surface, maybe 10 or 12 kilometres thick. That is the atmosphere of the Earth. That is it. Below that is life. Above it is nothing.
—JULIE PAYETTE, Canadian astronaut

Air has shaped life's evolutionary path in countless ways. Long ago birds exchanged forelimbs for wings and mastered a new realm. Many insects seem almost to be made of air—a bloom of mayflies hatching on a stream, butterflies drifting on invisible air currents, a buzzing, dancing cloud of microscopic gnats. Plants fling their seeds into the wind and attract pollinators by perfuming the air. Countless life forms exploit the air for the crucial ceremonies of their existence—broadcasting sounds, scents and molecules such as pheromones to attract mates, warn of danger or locate their young. Without air, we would not hear a child's whisper, or smell the sweet scent of honeysuckle, or feel the vibrations of a distant train. The simple act of one molecule of air bumping up against another and then another in an invisible wave allows us to sense our world into being. Without air, the canvas on which we paint an image of our world would lack depth and dimension. But most of all, the need for air has minutely shaped the physiology of all aerobic creatures. If you need reminding, simply try to stop breathing. You will quickly find that you have no choice in the matter. Within a few seconds, your body will *demand* air; within a minute, blood vessels in your head will bulge, your heart will pound, and your chest will heave with silent screams for air. We are more than just air breathers; we are creatures made for and by the substance we need every minute of our lives. And just as air has shaped and sustained living beings, so living beings created and still sustain the air.

We Are the Air

In North America the history of settlement and conquest has created a powerful myth of the primacy of the individual, free to act and move as an independent entity. But from a biological point of view, this myth is a mistaken and dangerous version of reality. We are not completely independent and autonomous; when we look carefully at the interactions at every level between our bodies and the element that surrounds us, we see how completely we are embedded in air, all of us caught together in the same matrix.

Air is a physical substance; it embraces us so intimately that we cannot say where we leave off and air begins. Inside as well as outside we are minutely designed for the central activity of our existence—drawing the atmosphere into the centre of our being, deep into the moist, delicate membranous labyrinth within our chests, and putting it to use.

Breathing is controlled by the oldest part of the brain, the respiratory centre of the brain stem, a relic that originated before the dawn of consciousness. Breathing is such a vital act that it has never been given over to the control of the later arrival—the conscious brain. Automatically, whether we are awake or asleep, that ancient link to our evolutionary past commands every single breath. From a newborn's forty breaths a minute, our inhalations

What You Can't See Might Kill You

Within the atmosphere both the proportion and the constituents of the air are crucial. In the sixteenth century, Spaniards invading the mountainous lands of the Incas encountered a mysterious disease. The "sickness of the Andes," as they called it, had something to do with the air, according to Father José de Acosta in his *Natural and Moral History* (1590):

> I persuade myself, that the element of the air is there so subtle and delicate, as is not proportionable with the breathing of man, which requires a more gross and temperate air, and I believe it is the cause that doth so much alter the stomach and trouble all the disposition.

It would be two centuries before the cause of the sickness was discovered—lack of oxygen. Altitude causes one kind of air-quality problem. Depth causes another; as Jonathan Weiner describes:

slow to thirteen to seventeen in our later years but can escalate to eighty during vigorous exercise, all without a conscious thought. If breathing is interrupted, most people suffer irreversible brain damage after two or three minutes without air and the finality of death within four or five minutes.

Our bodies possess an extraordinary number of built-in safety measures, fine-tuned to obtain just the right amount of air. In the aorta and carotid arteries, oxygen chemoreceptors constantly monitor the level of oxygen in the blood. When oxygen levels fall, the receptors send out impulses to the muscles of the diaphragm and ribs to increase the rate of breathing. Carbon dioxide or acid chemoreceptors respond to rising levels of acidity in the blood, which result when dissolved carbon dioxide forms

... most of the deadly gases in mines—carbon dioxide, carbon monoxide, methane, hydrogen—are odorless. Miners could pass out without warning, unless someone saw a worker sink to his knees farther down the shaft and had the presence of mind to shout the alarm: "Gas!" The only mine gas they could smell was hydrogen sulfide (they called it "stink damp") and even hydrogen sulfide was perverse. At very low concentrations, it smelled like rotten eggs but at higher, fatal concentrations it was completely odorless. Miners tried using mice, chickens, small dogs, pigeons, English sparrows, guinea pigs, rabbits. By trial and fatal error they settled on the canary: in the presence of carbon monoxide and stink damp, at least, the bird usually collapsed sooner than the miner holding the cage.

After World War II, more sophisticated gas detectors were invented and installed in mines. But even today, the blue cover of the Department of the Interior's latest safety manual on mine gases is decorated with a single yellow canary. □

carbonic acid. Again, the receptors send messages to the muscles of the diaphragm and ribs to increase respiration to eliminate the carbon dioxide.

There are also mechanoreceptors guarding the airways and lungs. In the lungs, stretch receptors detect pulmonary inflation. When you take a breath, the receptors send a signal that regulates the length of time before your next breath. Additional receptors coordinate breathing with levels of muscular activity, and other nerve centres regulate respiration when you are anxious, in pain, sneezing or yawning. You can override your unconscious control of breathing by deliberately holding your breath. But soon the rise in blood-borne carbon dioxide forces you to take a breath. Since the window of survival is only a few short minutes, your body has evolved a host of strategies to ensure a steady supply of the substance it cannot do without.

Oxygen is the crucial component; it has the ability to combust by sharing its electrons with other elements. This process, known as oxidation, can be so rapid as to ignite a fire or can be imperceptibly slow, as when iron rusts, or at controlled rates, as when metabolism takes place in living organisms. In a cell, oxygen breaks down molecules such as carbohydrates and fats, releasing energy in the form of heat. In the process, oxygen becomes a part of liberated carbon dioxide or other breakdown products. Oxygen lights the fires of life and keeps them burning.

The Path of a Breath

You could think of your upper body as one big air trap—an immensely complex mechanism designed to capture a lungful of air and put it to work. A breath begins when the diaphragm, a smooth muscle under the lungs, and other muscles between the ribs contract and pull the rib cage upward and outward to create a partial vacuum in the thoracic area. The weight of the atmosphere then forces air into your chest.

Although lungs have a volume averaging between 4.25 and 6 litres, lungs only draw in about 500 millilitres of air during rest. When you take a deep breath, this amount expands to 3 to 4 litres. Even when you exhale as deeply as possible, about a litre of air remains in your lungs.

Inhaled air rushes in through the nostrils, where it is filtered: large dust particles and foreign bodies are trapped by small hairs, which can expel these particles by inducing a sneeze, and smaller particles are filtered out by microscopic hairs along mucus-coated cartilaginous structures called conchae. Filtered, humidified

||||||||||||||||||||||||||||||||||||||||||||||||||||||||||||||||||||||||||||||||||||||||||||||||||||||||||||||||

Figure 2.1: The anatomical path of a breath of air.

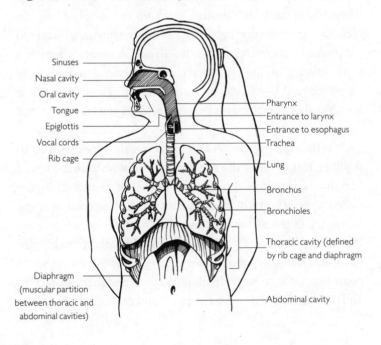

and warmed to body temperature, the air rushes along the roof of the nasal chamber and passes the olfactory organ, a small patch of mucous membrane that is rich in nerve endings. In pits and crevices of the olfactory organ, molecules in the air are sampled and the information is sent along a nerve way through the bony floor of the skull to the olfactory bulbs at the bottom of the brain. Although we seldom appreciate or even notice "pure, fresh air," we know instantly when there's smoke, perfume, rotting fish or lilacs in our surroundings. Even though we lack the exquisite olfactory discrimination of many other species, the air brings us detailed information about our surroundings. Smells stimulate appetite, arouse us, warn us, soothe us; often they trigger deep emotions and distant memories.

In the throat, air passes into the windpipe or trachea, which branches into two main bronchi that supply each lung (Figure 2.1). The bronchi divide into many smaller passages called segmental bronchi, which in turn break up into smaller bronchioles. The air passes through these steadily narrowing passages to culminate in grape-like sacs called alveoli. An average lung has 300 million alveoli, with a total cellular surface area the size of a tennis court! The alveoli are surrounded by capillaries, which are tiny tubes branching off from the arteries and carrying blood cells (Figure 2.2).

It is through the alveoli that air enters the bloodstream. To facilitate this process, the alveoli are lined with a three-layered film that is fifty times thinner than a sheet of airmail stationery. Called a surfactant, the film reduces the surface tension at the boundary of the air and the blood cells, thus encouraging diffusion of gases across it. The surfactant also adheres to inhaled particles until macrophages, our cellular garbage collectors, can come to cart them away. The air is literally fused to the surfactant, blurring the line where air leaves off and our cells begin as categories merge—gaseous and liquid, outside and inside—as the planet's atmosphere enters our bloodstream.

On average, we have about 5 litres of blood containing about 5 million red blood cells per millilitre (that's a total of 25 billion red blood cells). When we're relaxed, it takes about a minute for the 5 litres to complete the round-trip circuit from the heart to the lungs to the rest of the body and back to the heart. During exercise, this rate may accelerate sixfold.

There may be as many as 350 million hemoglobin molecules packed into every one of the 25 billion red blood cells. Hemoglobin

|||||||||||||||||||||||||||||||||||||||||||||||||||||||||||||||||||||||||||||||||||||||||||||||||||||||||

Figure 2.2: The detailed anatomy of an alveolus.

Figures 2.1 and 2.2 adapted from Cecie Starr and Ralph Taggart, *Biology: The Unity and Diversity of Life*, 6th ed. (Belmont, CA: Wadsworth, 1992), fig. 40.8.

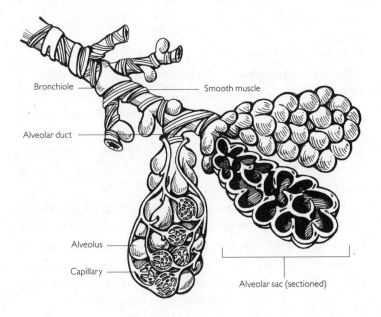

binds to carbon dioxide or oxygen molecules, transporting them to and from the lungs. Each hemoglobin molecule is capable of carrying four molecules at a time, so, in an average volume of blood, there are 35×10^{18} (that's 35 followed by 18 zeros) sites for oxygen or carbon dioxide. When the inhaled air comes into contact with the alveoli deep in the lungs, oxygen instantly diffuses across the membranes to attach to the hemoglobin, while carbon dioxide is exhaled into the atmosphere. Loaded with oxygen, the blood turns bright red and transports the vital fuel throughout the body to cells in need of energy.

When we are doing physical work or exercising, we draw on our energy reserves and may need up to 2.5 litres of oxygen every minute. As our exertion creates more carbon dioxide, it is released into the bloodstream, where it acts on the brain to stimulate more rapid breathing. That increases the intake of oxygen, and the heart is also stimulated to beat faster, thereby increasing the rate at which blood cells release carbon dioxide and pick up the oxygen in the lungs.

The Universal Glue

Since the entire gaseous contents of alveoli are not exhausted at each breath (when we are resting, only a tenth of the air is expelled), the air that remains keeps the alveolar sacs inflated and prevents them from collapsing. So air always resides within us and is as much a part of our bodies as any tissue or organ. We are a part of the air, which in turn is a part of all green plants and every other breathing creature.

After sharing space in a room with others, try a very simplified thought exercise. If you multiply the volume of air (in litres) in the room by Avogadro's constant (the number of atoms in 1 mole of substance: 6.022×10^{23}), you will get an estimate of the number of atoms in the air in that room. (Assume the air is always mixed completely.) Then divide the number of atoms in the air by the volume of air inhaled times the number of breaths per minute times the time spent in the room times the rate at which oxygen and carbon dioxide diffuse across lung cell membranes. Even the crudest calculation reveals that each of us very quickly absorbs atoms into our bodies that were once an integral part of everyone else in the room, and vice versa.

The eminent Harvard astronomer Harlow Shapley once performed another thought exercise about air. He pointed out that while 99 per cent of the air we breathe is highly active oxygen and mildly reactive nitrogen, about 1 per cent is made up of argon, an inert gas. Because it is inert, it is breathed in and out without becoming a part of our bodies or entering into metabolic transformations. Shapley calculated that each breath contains about 30,000,000,000,000,000,000, or 3.0×10^{19}, atoms of argon plus quintillions of molecules of carbon dioxide. Suppose you exhale a single breath and follow those argon atoms. Within minutes, they have diffused through the air far beyond the spot where they were released, travelling into the neighbourhood. After a year, those argon atoms have been mixed up in the atmosphere and spread around the planet in such a way that each breath you take includes at least fifteen atoms of argon released in that one breath a year earlier! All people over the age of twenty have taken at least 100 million breaths and have inhaled argon atoms that were emitted in the first breath of every child born in the world a year before! According to Shapley:

> Your next breath will contain more than 400,000 of the argon atoms that Ghandi breathed in his long life. Argon atoms are here from the conversations at the Last Supper, from the arguments of diplomats at Yalta, and from the recitations of the classic poets. We have argon from the sighs and pledges of ancient

lovers, from the battle cries at Waterloo, even from last year's argonic output by the writer of these lines, who personally has had already more than 300 million breathing experiences.

Air exits your nose to go right up your neighbour's nose. In everyday life we absorb atoms from the air that were once a part of birds and trees and snakes and worms, because all aerobic forms of life share that same air (aquatic life also exchanges gases that dissolve back and forth at the interface between air and water).

Coping With Hypoxia

Whales, dolphins, seals and other marine mammals seem so at ease in the water that it is easy to forget they are truly creatures of the air. Like us, they can drown. To cope with the fact that their time at the ocean-air interface is fleeting, marine mammals, particularly the deepest-diving ones, have adapted strategies to conserve the oxygen in each precious breath, thus thwarting the metabolic risks of hypoxia (low oxygen).

Unlike terrestrial creatures, marine mammals breathe consciously, controlling every breath. They take full advantage of those few moments at the surface, forcefully exhaling waste gases and deeply inhaling new, oxygen-rich air. Fin whales, for example, exchange up to 90 per cent of their lung volume with each powerful breath.

Elephant seals are diving wunderkinds. They can dive over a kilometre deep for as long as an hour and, remarkably, spend just three minutes or so on the surface between dives. Elephant seals do not hold their breath as they dive (just imagine trying to force an inflated beach ball deep into the water); instead, their lungs, alveoli and trachea collapse. Only 5 per cent of the air remains in their lungs. Instead, most of the oxygen is ferried into the blood

Air is not a vacuum or empty space—it's a physical substance that embeds within us all and perfuses through us. It is constantly changing as life and geophysical forces add and subtract constituents to the composition of air, and yet over vast stretches of time the basic composition of air has remained in dynamic equilibrium. The longer each of us lives, the greater the likelihood that we will absorb atoms that were once part of Joan of Arc and Jesus Christ, of Neanderthal people and woolly mammoths. As we have breathed in our forebears, so our grandchildren

and muscles, where a large blood volume (up to 20 per cent higher than humans) and huge stores of myoglobin (three times as much as humans) increase the efficient use of this most precious resource.

To further conserve oxygen, elephant seals slow their metabolic rate during a deep dive. Blood is shunted from the extremities to the central organs, such as the heart and brain, and their heart rate slows from 110 to 120 beats per minute, to 20 to 50 beats per minute.

Humans living at high altitude also have adaptations to oxygen-poor conditions. As altitude increases, the air becomes thinner: there are fewer oxygen molecules in each breath of air. To compensate, inhabitants of the high plateaus of the Andes have blood with a greater percentage of hemoglobin and are thus able to hold more oxygen. Interestingly, this strategy did not evolve universally in all high-altitude populations. People of the Himalayas and Tibetan Plateau do not have the same elevated hemoglobin levels (except at extreme altitudes of 4000 metres or more); instead, they compensate for low oxygen by taking more breaths per minute. □

and their grandchildren will take us in with their breath. We are bound up inseparably with the past and the future by the spirit we share.

Every breath is a sacrament, an affirmation of our connection with all other living things, a renewal of our link with our ancestors and a contribution to generations yet to come. Our breath is a part of life's breath, the ocean of air that envelops Earth. Unique in the solar system, air is both the creator and the creation of life itself.

The Origin of Air

The scientific explanation of the origin of air is as awesome as any other origin story, as majestic in its sweep of time and in the scale of its events. Scientists believe that in the aftermath of the Big Bang that created the universe, immense clouds of swirling gases cooled and condensed into clots of matter drawn together by the attractive forces of their gravity. The contraction of matter heated the core of massive bodies until atoms overcame the electrostatic repulsive forces that keep them apart. Hydrogen atoms were drawn together so tightly that they pierced the powerful repellent forces of surrounding electrons. The nuclei of these hydrogen atoms fused and formed helium while liberating energy, thus igniting the thermonuclear furnaces of stars that lit up the heavens throughout the expanding universe.

Ten billion years after the Big Bang, a star—our sun—was born in the Milky Way galaxy. Circling that star were clouds of gas that coalesced into smaller bodies called planets. One of the planets was Earth, which was born some 4.6 billion years ago out of an aggregation of dust and meteorites. It continued to grow for millions of years, absorbing any bodies that came its way, sweeping up cosmic dust as it orbited the sun. Hydrogen and helium in the original mantle surrounding Earth were too light to be held by the planet's gravity and escaped into space. They left behind a primordial atmosphere that is thought to have been 98 per cent carbon dioxide, 1.9 per cent nitrogen and 0.1 per cent argon.

As Earth cooled, it became geologically active—volcanoes spewed forth vast quantities of lava, ash and gases. Then, as now, most of the gaseous emissions were water vapour, carbon dioxide and compounds of sulphur, nitrogen and chlorine. In addition, molecules of methane and ammonia formed from the elements in the gases. But there was no free, or pure, oxygen, which is necessary to sustain animal life. The gases that did exist, known as greenhouse gases, formed an atmospheric envelope that was transparent to the sun's rays. These rays penetrated to the planet's surface as shorter wavelengths of light. The radiation from these wavelengths was reflected back toward space, where the greenhouse gases behaved like the glass of a greenhouse and trapped longer wavelengths, such as infrared, holding heat like a blanket, raising the surface temperature of Earth (Figure 2.3).

IIIIIIIIIIIIIIIIIIIIIIIIIIIIIIIIIIIIIIIIIIIIIIIIIIIIIIIIIIIIIIIIIIIIIIIIIIIIIIIIIIIIIIIIIIIIIIIIIIIIIIIIII

Figure 2.3: The absorption of heat by greenhouse gases.

Adapted from Jeremy Leggett, ed., *Global Warming: The Greenpeace Report* (Oxford: Oxford University Press, 1990), p. 15.

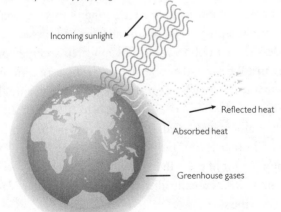

Incoming sunlight

Reflected heat

Absorbed heat

Greenhouse gases

Because Earth's atmosphere was rich in greenhouse gases and was twenty to thirty times as dense as today, the surface of Earth heated up to a temperature of perhaps 85° to 110°C. When volcanic activity subsided, the atmosphere gradually cooled enough for water vapour to condense into clouds, which eventually rained onto the land. Thus began the water cycle, the continuous process of condensation-precipitation-evaporation that is so crucial to life. There was no soil, and over hundreds of millions of years, water accumulated in rivers, lakes and the oceans. Over eons, infinitesimally small quantities of salts and elements leached from rock and accumulated in the oceans.

The Interaction of Life and Air

Atoms and simple molecules (combinations of atoms) in the atmosphere wafted across the surface of the ocean and dissolved into the waters. Atoms continued to accumulate as rivers flowed across the land and into the seas. Eventually this rich mix of atoms and molecules interacted to form even more complex structures that were to become the building blocks of the large compounds in living cells—nucleic acids, proteins, lipids and carbohydrates—that transmit hereditary information, carry out metabolic reactions and form the cellular structures of all organisms. In this prebiotic environment, conditions were accumulating that made it possible for life to catch fire and spread.

In less than a billion years, life appeared spontaneously in the oceans. We can only speculate on the bizarre experimental forms that the first cells, or protocells, took, but eventually one cell acquired the properties that enabled it to succeed where countless others had failed. It was able to win the competition against all other forms and to persist by replicating. That cell was the ancestor of every living thing that exists today—one cell whose progeny eventually filled the oceans, covered the land and soared into the skies.

Were we to invent time travel and go back 2 billion years, the Earth would be alive with organisms, but none would be visible to the human eye. It was a microbial world; bacteria were the only forms of life on Earth. They survived in the world's oceans, in what was essentially an expansive pool of chemical experimentation. Their world was anoxic (without oxygen), and these bacteria extracted energy from the chemicals around them, such as hydrogen sulphide or carbon dioxide. Then, about 2.5 billion years ago, a group of micro-organisms called cyanobacteria developed a way to capture the energy of photons streaming to Earth from the sun and convert it into high-energy chemical bonds that could be stored in molecules and called up when needed.

The language of life is chemistry.
—LYNN MARGULIS, quoted in John Brockman, *The Third Culture*

This ability to convert the sun's energy into food, a process called photosynthesis, transformed the nature of life on Earth. Before cyanobacteria, life had to scrounge for energy, extracting it from deep-sea vents, volcanoes and chemical bonds. Sunlight, though, was limitless. With photosynthesis, energy became abundantly accessible. The key to photosynthesis is the pigment chlorophyll, which traps photons and transfers the energy to sites where it can be used to make sugar. In photosynthesis, cyanobacteria (and, later, algae and plants) added the captured energy of light to a mix of carbon dioxide and water to produce glucose, a simple sugar molecule with a backbone of carbon and oxygen atoms. And incidentally, as cyanobacteria performed their solar-powered chemistry, life began the long process of modifying the atmosphere that continues to this day.

For every six molecules of carbon dioxide transformed by photosynthesis into a molecule of sugar, six molecules of oxygen (O_2) are released. For 100 million years, oxygen was released by cyanobacteria into the ocean. There it reacted with iron in solution to form iron oxides. Essentially, the ocean began to rust. The iron oxides precipitated out and fell to the ocean floor. Today, these sediments are the source of much of the world's iron ore.

The geological record shows that about 2 billion years ago terrestrial rocks containing iron began to rust, too. These "red bands" provide a clue to the approximate time when the ocean was finally saturated with oxygen; it was the atmosphere's turn to receive this chemical gift. Little by little, the balance of gases shifted imperceptibly but inexorably toward the oxygen-rich atmosphere we know today. And as life altered the chemical composition of our atmosphere, it created new opportunities for life. The Oxygen Revolution had begun.

Cyanobacteria transformed the chemistry of Earth and primed the environment for new forms of life to evolve. Eventually, new life did appear when one bacterium invaded another, actually penetrating a host's membrane to lodge inside the

Window on the Past

Shark Bay, Australia, provides a glimpse into Earth's history, where we can visualize our planet as it might have been a few billion years ago. The cauliflower-like boulders in the ocean shallows are called stromatolites. Their construction crew? Cyanobacteria.

Stromatolites formed as mats of cyanobacteria were covered with sand and mud. Since cyanobacteria need light, they move to the surface and continue to build upward, forming a layered living rock of bacteria and mud.

From these unassuming clumps of rock, cyanobacteria capitalized on the abundant resources around them. Using sunlight to power a reaction that released the oxygen from water, they created an atmosphere for life on Earth. ☐

protoplasm. Although it was likely a simple survival strategy—a search for food or protection from a predator, perhaps—this invasion proved to be a turning point in the history of life. Rather than creating an antagonistic union, both host and invader discovered mutual advantages from this relationship. The new whole was indeed greater than the sum of its parts: co-operation prevailed over competition.

From this partnership between two bacteria, a new type of cell evolved—one with greater complexity, with a nucleus and organelles—a cell that facilitated the origin of more complex life. Although a controversial idea even into the 1980s, it is widely accepted today that cell mitochondria evolved when one bacterium merged with another. (The mitochondrion is the powerhouse of cells, using oxygen and chemicals to create fuel.) The chloroplasts that enable plants to photosynthesize are the result of a similar symbiotic partnership. The DNA in a chloroplast bears little resemblance to that in the cell's nucleus but is striking in its similarity to the DNA of cyanobacteria.

From a world dominated by bacteria, the fortuitous merging of early microbes in the Earth's oceans made more complex multicellular aggregates—algae, fungi, plants and animals—possible. At first, organisms continued to evolve in the cradle of the ocean, and then on the moist sand, clay and dust at its margins. It wasn't until about 475 million years ago that some green plants began to colonize land, and, in doing so, continued to transform the atmosphere.

The invasion of land by plants led to a blossoming of life—a massive increase in both the diversity and the numbers of living things. Over an immense period of time, plants were established as dominant life forms in the oceans and on land, thereby enabling the evolution of animals called herbivores that could exploit plants. Grazing animals incorporated the breakdown products of plant molecules into their own body structures, and the carnivores that fed on them incorporated those molecular remnants into their bodies. For eons, generations of plants and animals flourished and died; as their carcasses piled up and decomposed, the molecules that once formed them leaked back into the soil. Eventually, those chains of carbon became the "fossil fuels"—peat, coal, petroleum and natural gas—that we exploit today for the energy still stored in their chemical bonds.

Thus, air and life have constantly interacted, altering each other in a dynamic process of perpetual change. While carbon dioxide was removed from the air and water and sequestered in calcium carbonate shells (which eventually formed limestone deposits) or fossil fuels, oxygen was liberated by photosynthesis. Billions of years ago, carbon dioxide dominated the atmosphere. Gradually, living organisms transformed that atmosphere as the dominant constituents of air became nitrogen (78.08 per cent), oxygen (20.95 per cent) and argon (0.93 per cent) (see Table 2.1). Only when the atmosphere was enriched with oxygen could life as we know it evolve and flourish.

An Atmosphere for Life

If the Earth were reduced to the size of a basketball, the part of the atmosphere where weather occurs and all organisms live would be thinner than the finest paper. In that thin coating of gritty slime, life has taken hold and flourished. The balance of atmospheric content and temperature has been crucial to that success, as can be seen by comparing Earth with its neighbouring planets, Venus and Mars. The Venusian atmosphere, one hundred times as dense as Earth's, is composed primarily of the potent greenhouse gases water vapour and carbon dioxide. As a result, the average surface temperature of Venus is an inhospitable 460°C. In contrast, Mars has an atmosphere that is 95.3 per cent

II

Table 2.1:
Relative Proportions of Gases in the Lower Atmosphere

GAS	% BY VOLUME	PARTS PER MILLION
Nitrogen	78.08	780,840.0
Oxygen	20.95	209,460.0
Argon	0.93	9,340.0
Carbon dioxide	0.035	350.0
Neon	0.0018	18.0
Helium	0.00052	5.2
Methane	0.00014	1.4
Krypton	0.00010	1.0
Nitrous oxide	0.00005	0.5
Hydrogen	0.00005	0.5
Xenon	0.000009	0.09
Ozone	0.000007	0.07

carbon dioxide, but the total volume of the atmosphere is only 0.6 per cent of Earth's at sea level. The Martian atmosphere is too diffuse to trap heat, so its average surface temperature is a frigid −53°C. We swing through space together like an object lesson—the fiery furnace, the cold chunk of rock and the bright, living world of home.

Life has flourished in this planet's ephemeral mantle of air, which has a total density of 5.1×10^{15} tonnes, less than one-millionth of the Earth's entire mass. The atmosphere extends 2400 kilometres above the Earth's surface, yet all but 1 per cent of it is within 30 kilometres of the ground, 5 million billion tonnes of air squeezed more and more densely as it gets closer to the planet's surface. In the lowest layer it presses down with a force of 1 kilogram per square centimetre—the pressure that we, like the other life forms we share the surface with, are adapted to and cannot do without.

Our relationship with air pressure at the Earth's surface is so fine-tuned that we are endangered not long after leaving its protective, perfectly pressurized cloak. Thanks largely to the efforts of pioneering mountaineers and balloonists, we know it doesn't take too much altitude before we are out of our comfort zone when it comes to atmospheric pressure and the availability of oxygen. Even the most earthbound of us have felt one of the first signs of altitude shifts—aching eardrums. On the ground (at whatever altitude above sea level we live), the air pressure on both sides of the eardrum is equalized. As air pressure drops with increasing altitude, the air with higher pressure inside the ear pushes the eardrum outward, resulting in severe pain unless the pressure can be released. (The reverse happens when a diver descends—the eardrum is pushed inward by water pressure.)

As we ascend, catching our breath also becomes increasingly difficult; there is less oxygen available per breath. As well, high altitude and lowered air pressure can cause fluid to leak from capillaries and pool in the lungs and the brain. Eventually, we cannot get enough oxygen to supply our needs, and without descending or using a breathing apparatus, death comes quickly. The ability to bottle air at its pressure at Earth's surface expanded our habitat—albeit for brief forays—opening the skies and the seas to human explorers.

We need to explore the complex layer cake of gases that presses down on our heads so that we can understand what happens there and why. In Europe in the Middle Ages the view from Earth was pictured as a series of crystalline spheres that revolved

Figure 2.4: The layers within the atmosphere.

The dotted line indicates temperature in the lower atmosphere.

Adapted from *Science Desk Reference* (New York: Macmillan, 1995).

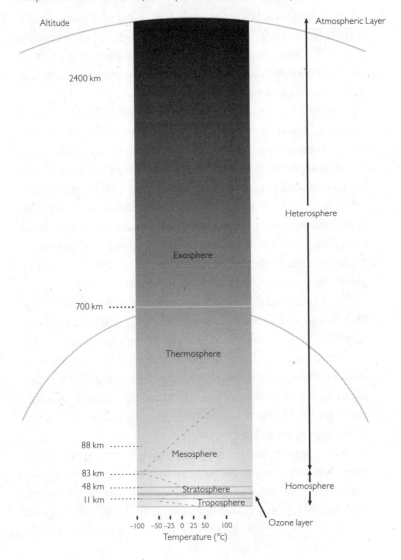

constantly within each other, carrying the stars. As they moved they sang, creating vast heavenly harmonies—the music of the spheres. Today we can adapt this conception to a vision that is radically different but no less remarkable—the atmosphere and all its smaller spheres that wrap this planet (Figure 2.4). They can be imagined as a set of zones, each containing various materials with different properties, moving within each other. Although they do not sing, these regions shine, reflect, protect, warm and cool—they are life's habitat. Sheltered within our spheres, we are the ones who can look up and sing.

Viewed from Earth's surface, the atmosphere seems homogeneous, constantly mixed by winds and convection. In fact, the first 83 kilometres above Earth is called the homosphere because the air is kept evenly mixed. Half of the mass of the atmosphere lies less than 6 kilometres above Earth. Even this diffuse veil has differences that reflect variations in solar radiation, heating by Earth's surface, gravity, evaporation and incoming cosmic radiation. Thus, for example, gravity holds the heavier elements closer to the ground, while lighter ones, such as helium, are found in greater relative abundance at extreme altitudes.

The lowest level of the homosphere is the troposphere, where life is found and weather occurs. It averages 11 kilometres above Earth but is 8 kilometres above the poles and 16 kilometres above the equator. Above the troposphere is the stratosphere (11 to 48 kilometres above Earth), where gases become thinner; this region contains the ozone layer, between 16 and 48 kilometres above Earth. Above the stratosphere lies the mesosphere, 48 to 88 kilometres above Earth.

Down here in the troposphere, where we live, there are different levels of atmospheric action, many of them overlapping and interacting. High-speed winds separate the flow of warm air from the equatorial area and cold air from the polar regions.

These winds are called the jet streams and range from 7620 to 13,716 metres above Earth. In general, the jet streams move from west to east in both the Northern and Southern Hemispheres, but they can temporarily curve north or south. Westerly winds in the Northern Hemisphere often follow the path of the jet streams above.

Most of the time we are unaware of the large movements of the Earth and its atmosphere. But sometimes we are directly in touch with the engine that drives life. Anyone who flies round trip between Vancouver and Toronto or between San Francisco and New York will be struck by the fact that it takes over five hours to fly west but less than four hours going from west to east along the same route. The reason for the difference is the headwinds that sweep from west to east as the atmosphere is pulled by the rotation of Earth. Such air currents continually mix contributions from different parts of the globe—water vapour from the oceans and rainforests of the planet, dust from deserts and emissions from human industrial centres—carrying and spreading debris like immense rivers in the air.

The troposphere is constantly stirred and mixed by convection currents that arise from differences in temperature between air, land and sea; from mountain ranges, weather and moisture; from marine algal blooms and forest transpiration. Although the complexity of these perturbations makes local weather hard to predict, global wind patterns can nevertheless be mapped and understood. Air between the Northern and Southern Hemispheres doesn't mix much—as the air warms around the equator, air currents are generated on either side that are separated by the windless doldrums between.

The Grasshopper Effect

Glaciers high in the Rocky Mountains, vast Arctic expanses of tundra, snow, ice and sea: the apex of Canada and the top of the world. Far from industry, automobiles and major cities, these are the last places you'd imagine to find disturbing levels of contamination. But sadly, these cold, remote areas are the unwitting recipients of some of the world's most persistent pollutants.

The late David Schindler took ice cores from Snow Dome Glacier, high in the Canadian Rockies, for years. As a new layer of ice forms each year, an annual record of atmospheric pollutants—hitchhikers on the world's winds—is preserved. And the record is an alarming one—pesticides, PCBs, and other persistent organic pollutants are trapped, leaving an icebound account of our chemical creations.

The journey of these pollutants to the ice may have been a long one. It likely began in a warmer climate, where the chemicals vaporized and joined the long-range transport system of atmospheric winds. They were carried along until they hit colder air, where they condensed and fell to Earth, piggybacking on rain or snow. This cycle of warming (vaporization), cooling and precipitation might have happened over and over before the pollutants finally stopped their globe-trotting, in higher, colder and more northerly climes. Since the journey occurred in great northward "jumps," this phenomenon is dubbed the Grasshopper Effect.

The traditional marine-based diet of Indigenous Peoples of the circumpolar North is considered one of the world's healthiest. But now, this diet, which not only sustains them but also defines their cultures, threatens their health. As pollutants precipitate out in the water or soil, they enter the bottom link of the marine food chain. These toxins are passed on, accumulated and concentrated, from plankton to snail to fish, until they reach the end of the

Arctic food chains—northern people and top predators such as polar bears, seals and whales.

> Breast milk is supposed to be a gift. It isn't supposed to be poison.
> —ERIC DEWAILLY, quoted in Marla Cone, *Silent Snow*

Although all of the world's people carry some traces of toxic chemicals, it is a sad irony that people who live in one of the least industrialized places on Earth carry in their bodies a toxic record of the pesticides and industrial compounds that we have unleashed on the planet. Women in Nunavik, on the shores of Hudson Bay, have seven times as many PCBs in their breast milk than women in some of the largest cities in Canada; the average levels of PCBs and mercury in women's breast milk in Greenland is twenty to fifty times higher than in the urban United States and Europe; increased levels of toxic contaminants in polar bears are leading to changes in their hormone and immune systems. The list goes on.

In more dramatic fashion, when fire broke out in the Chernobyl nuclear plant in Ukraine on April 26, 1986, it was Swedish scientists who first alerted the world that something catastrophic had happened in the Soviet Union. The radioisotopes liberated from Chernobyl's nuclear cauldron poured out of the wreckage into the atmosphere. As they blew across Scandinavia, instruments detected a huge spike of radioactivity. Like identification tags, those radioisotopes moved around the Northern Hemisphere to reveal the path of air from Ukraine. Far away in Wales, radioactive fallout was so intense that sheep became contaminated and were banned from sales. A year later, the ban was still in force.

Chernobyl and the contamination of the Arctic graphically illustrate the global nature of air. Air is not a national or a local resource but a global commons into which we contribute our wastes and from which we draw air to fuel our bodies. □

The Atmosphere as Radiation Shield

The atmosphere has played another crucial role in the development and survival of life on Earth—that of a radiation shield. In addition to visible light, the planet is constantly bombarded by invisible, short-wavelength ultraviolet light from the sun. Nucleic acids, the hereditary material of living organisms, are particularly susceptible to this wavelength of light. When ultraviolet light hits a molecule of DNA, specific parts of the molecule absorb energy from the light, thereby inducing chemical modification and reaction. Although "repair mechanisms" have evolved to deal with ultraviolet-induced DNA damage, not all damage is repaired and genetic alterations can result. Because life has evolved over long periods of time, almost any alteration of the genetic material is likely to upset the balance of finely honed gene-mediated reactions in the cell and will prove deleterious rather than beneficial.

In the atmosphere, when an oxygen molecule made up of two oxygen atoms is struck by ultraviolet light, it is split apart into two free radicals, or highly reactive atoms. An oxygen free radical can react with an oxygen molecule made up of two atoms to create an oxygen molecule of three atoms, or ozone. Thus, oxygen in the atmosphere is constantly being broken down and recombined into ozone. Thirty kilometres above ground, in a layer as thick as a sheet of newspaper, is the zone where ozone is formed and degraded. The zone is called the ozone layer. Since ozone is a form of oxygen, it has the ability to trap the energy of ultraviolet light; thus, the ozone layer filters out much of the ultraviolet light before it can strike Earth.

Once a photograph of Earth, taken from the outside, is available, an idea, as powerful as any in history, will let loose.

—FRED HOYLE, quoted in E. Goldsmith et al., *Imperiled Planet*

This chemical interaction is part of a splendid symmetry, part of the ancient reciprocal relationship between life and the atmosphere, each creating, adjusting to, modifying and protecting the other over the millennia. In the context of planetary and biological evolution it has been an extraordinary collaboration—and it may have reached an equilibrium. Viewed over the long sweep of geological change, Earth's atmosphere has been in constant flux, affecting and being affected by life itself. But for several hundred thousand years, through ice ages and interglacial periods, the atmosphere has remained relatively static. Oxygen has hovered around 21 per cent of the atmosphere, a propitious level, since 25 per cent oxygen could well ignite the atmosphere; if the atmosphere contained only 15 per cent oxygen, it would be lethal to life. Carbon dioxide and water vapour, which are greenhouse gases and are also conducive to photosynthesis, have kept the surface temperature of the planet within the limits of a comfortable 7°C variation over the past 3 million years. These proportions and relationships may be the harmonies we hear, nestled in our hollow on Earth, sipping the air. Moving and turning above us, around us, within us is the invisible element that first animated the planet—the breath of life. As someone said, it wouldn't be a fish that first discovered water. And so we had to leave home to understand the full meaning of what air and life have created together.

When we look into the sky it seems to us to be endless... We think without consideration about the boundless ocean of air, and then you sit aboard a spacecraft, you tear away from Earth and within ten minutes, you have been carried straight out of the layer of air, and beyond there is nothing! Beyond the air there is only emptiness, coldness, darkness. The "boundless" blue sky, the ocean which gives us breath and protects us from the endless black and death, is but an infinitesimally thin film.

How dangerous it is to threaten even the smallest part of this gossamer covering, this conserver of life.

—VLADIMIR SHATALOV, quoted in K.W. Kelley, ed., *The Home Planet*

Air for All Life

Invisible and indivisible, air is a place without borders or owners, shared by all life on Earth. It is the rightful inheritance of all future generations, the matrix that has shaped the course of evolution. Air binds us all together as a single living entity extending through time and space. Each one of us, past, present and future, needs air every minute of every day we live, in the proportions and the purity our bodies are adapted to.

Air quality depends now, as always, on the dynamic interaction between life and the atmosphere—on what goes in and what is taken out. Over vast periods of time, as the proportion of gases in the atmosphere has changed, so have the kinds of organisms that thrive. But now change is happening very rapidly, as human technology pushes the system. The exhalations of our machines are adding to and altering the constituents of air.

For many hunter-gatherer groups, protecting the spirits of the animals they kill is a sacred responsibility. Performing the proper rituals, killing no more than they need and wasting no part of their prey are all ways in which they express their gratitude and acknowledge their dependence on these animals. In the same way, but even more deeply, we need to acknowledge our responsibility to protect the air we breathe.

How Earth retains its ability to keep air fit for life remains poorly understood by scientists—like so much else. But the priorities are obvious enough. The planet's layer of photosynthesizing organisms is crucial in contributing oxygen to the atmosphere. Automobiles are a major source of carbon dioxide.

Since the Industrial Revolution, levels of atmospheric carbon dioxide have been rising steadily and could more than double by 2100 if there is no effort to limit our dependence on burning fossil fuels. A global agreement to protect and augment forests and marine plants is an essential beginning, affirming the priority of air as the bottom line for life—our life, all life, always. But such an agreement would do little to solve the long-term problem without worldwide reductions in emissions from human technology. And that means weaning ourselves from our fatal addiction to fossil fuels.

From our first cry announcing our arrival on Earth to our very last sigh at the moment of death, our need for air is absolute. Every breath is a sacrament, an essential ritual. As we imbibe this sacred element, we are physically linked to all of our present biological relatives, countless generations that have preceded us and those that will follow. Our fate is bound to that of the planet by the gaseous exhausts of fires, volcanoes and human-made machines and industry.

Once we have restored the breath of life to its rightful primacy—the first above all other human rights and responsibilities, the reference point from which all decisions flow—we can start to work in the long term to revive an ancient equilibrium. Using nature as our touchstone, we can play our part once more in life's long collaboration with the air.

3

The Oceans Flowing
Through Our Veins

And with water we have made all living things.

—KORAN 21:30

He sendeth the springs into the valleys, which run among the hills.
They give drink to every beast of the field; the wild asses quench their thirst.

—PSALM 104:10–11

HAD THE EARLIEST explorers of our solar system been transgalactic adventurers from another part of the universe, their first sight of this planet might have led them to name it Water. From space, you can see that ours is not the green planet but the blue planet, with its great oceans and its gossamer veil of clouds.

An astounding 70.8 per cent of Earth's surface is ocean; with an average depth of 3.73 kilometres, the oceans contain a total of 1.37 billion cubic kilometres of water. When inland seas, lakes, glaciers and polar ice caps are included, a total of 379.3 million square kilometres—74.35 per cent of the planet's surface—is covered by water. The land masses above the surface are just bumps. If the solid part of Earth were to be smoothed and levelled, a single ocean would wrap the entire globe to a depth of 2.7 kilometres.

Human beings are landlubbers on this watery planet, island people marooned on dry land, surrounded by and dependent on an alien element, an old home we left long ages ago and yet carry still within us. Water is the raw material of creation, the source of life. When the waters break, the child is born from them, just as the gods of old parted the dark, primeval ocean and fashioned the Earth, just as the first land creatures struggled up out of the tide.

Perhaps that is why water is at the heart of human ritual. Baptism, for example, often welcomes the child into the human family, washing away the past, marking a new start. The powerful symbolism of water—as transformation, purification, sharing—permeates our lives. Water flows through our memories: that sunlit swim in a creek, that wish made as a coin falls into a fountain, that first sigh of pleasure when tired feet are slipped into a cold mountain stream. Our literature is saturated with our uncertain relationship with this crucial substance—the water we come from, the water we cannot do without, the water that may drown us or flood away our world.

> *Full fathom five thy father lies;*
> *Of his bones are coral made:*
> *Those are pearls that were his eyes:*
> *Nothing of him that doth fade,*
> *But doth suffer a sea-change*
> *Into something rich and strange.*
>
> —WILLIAM SHAKESPEARE, *The Tempest*

The ocean—shifting, changeable, mysterious—has a powerful influence on human life and grips the human imagination. Rising and falling around Earth's shores, it moves to more than terrestrial rhythms. Pulled three ways, by Earth, the moon and the sun, the tides wax and wane day by day, month by month,

Hypersea: An Ocean Flowing Through the Land

Life may have evolved in the ocean—where it stayed for about 3.5 billion years—but in comparison, life's success on land has been spectacular and swift. Once plants started to colonize Earth's terrestrial margins, about 475 million years ago, the diversity of life expanded quickly. Today, it is estimated that there are two species on land for every one in the ocean. And all of this life is crammed into a space a fraction of the ocean's depth and breadth. What is it that caused this spectacular explosion of life?

We know that terrestrial organisms carry the ocean within them—in fluid-filled cells, for instance. This is how organisms were able to exploit the land, but Dianna and Mark McMenamin took the idea further. They suggest there is another "ocean" that flows through life on land. Whereas marine organisms are passively and individually bathed in fluid, terrestrial organisms are connected physically with other organisms, creating complex networks through which fluids can move. For example, parasites may travel through the blood of an animal, tree roots intertwine with the tiny threads, or hyphae, of a fungus that weave through the soil, and aphids drink the sap of a plant. The crux of the McMenamins' idea is that organisms on land became increasingly successful, pressured largely by the influence of fungi and parasites, and were able to diverge and colonize new habitats as they learned to exploit fluids. But they weren't using just the obvious fluids, such as surface water, but also the internal seas flowing in blood, sap and cellular fluid. These fluid connections resulted in another aquatic habitat on land, an internal sea flowing between living things. □

season by season, dancing to the rhythmic gyrations of planet, satellite and star. Somehow we have always known this and listened for un-Earthly messages in the motions of the sea. The ancient Greeks called the messenger Proteus the old man of the sea, herdsman of the ocean's seals. He saw the future and would tell you the truth about it—if you could catch him. Metamorphosis was his escape; changing his shape from lion to dragon to a stream of water, becoming a flame, a tree, he slipped through your fingers in a dizzying series of transformations. In the same way, the waters he represented are eternal shape-shifters, continually transforming themselves and the rest of the planet. They bring us a strange and ancient truth that is hard for us to grasp: a vision of the sources of life and its endless metamorphoses.

The Hydrologic Cycle

If air is the fuel, the spirit that animates all living things, water gives them body and substance. Water was absolutely necessary for life as we know it to have evolved. Life originated in the oceans, and the salty taste of our blood reminds us of our marine evolutionary birth. But we, like many other animals and plants, cannot live on salt water. Our lives are made possible by the hydrologic cycle, the miraculous process whereby salty water is transformed into fresh water by evaporation and is redistributed around the planet. Energy from the sun causes water to evaporate from the ocean as water vapour, which rises into the atmosphere and then falls back onto the land as precipitation. Water reaching Earth's surface as rain seeps into the ground or runs into rivers and lakes and eventually returns to the oceans (Figure 3.1).

The hydrologic cycle is crucial to all life, though sometimes we may wish it were not. "It's raining again!" is a common complaint in Seattle and Vancouver. "It poured for five days," a disappointed tourist grumbles in the Choco rainforest in Colombia, forgetting that the lush splendour he has come so far to experience was created and maintained by the rain he deplores. Rains bring life. Those working close to the land know this. Coastal fishers hold a collective breath if warm, dry weather lingers late into the fall. They anxiously watch as salmon, ripening with eggs and milt, mill in the estuaries waiting for the rains to come and the rivers to fill. Prairie farmers gazing gloomily upward into a cloudless sky understand the hydrologic cycle better than most. So do Indigenous people, whose rain dances beg the clouds to cry,

||||||||||||||||||||||||||||||||||||||||||||||||||||||||||||||||||||||||||||||||||||||||||||||||||||||||

Figure 3.1: The hydrologic cycle.

Adapted from Charles C. Plummer and David McGeary, *Physical Geology*, 5th ed. (W.M.C. Publishers, 1991), p. 234.

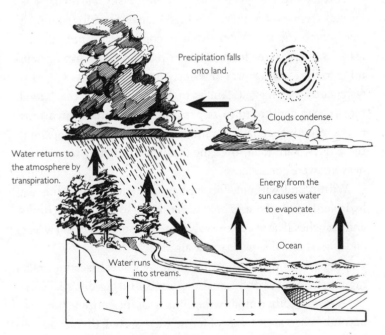

and cultures in other parts of the world, whose elaborate festivals entreat the gods to send the life-giving monsoons.

Living organisms are active participants in the hydrologic cycle, absorbing and filtering water and breathing it back into the atmosphere. Plants play a particularly important role through transpiration, or the loss of water through their leaves.

A forest is an intricate device for catching, holding, using and recycling water. You might liken it to a large sponge, except that it is far more complex. That tangle of tree roots snaking across the forest floor absorbs water while holding the soil so effectively that creeks don't flood and the water flowing in them is clean and clear after many days of rain. Millions of tonnes of water in rainforests are lifted from the soil and thrust back into the sky by transpiration. In essence, forests rain upward, into the heavens.

Forests also harvest moisture from the sky. Particularly in coastal areas, where there is a high incidence of fog and mist, trees comb the air. Water droplets condense on needles and leaves and fall to the forest floor, drip by precious drip contributing to the forest's stores of water. And trees' capacity to gather water can be significant. Studies in Australia showed that "cloud forests" can capture 10 to 25 per cent more water from atmospheric vapour than falls as rain. Imagine what would happen if these trees, with their towering canopies stretching toward the heavens, were gone.

Within and without, forests help regulate climate and the hydrologic cycle. Forests hold water in the soil, in roots, trunks and branches. The water is slowly meted out over days and weeks, and any excess is returned to the air.

Large areas of forest create their own local weather, raining on themselves and remaining moist during dry spells. At the same time, they modify the climate of the entire region and beyond. When large tracts of forest are removed, the barren soil hardens, causing rain to evaporate or run off rapidly.

The Life of Aqua, a Molecule of Water

Suppose we were to follow a single molecule of water vented from an active volcano on a Hawaiian island. We'll call this molecule Aqua. Liberated with a mix of other gases from deep within the planet, Aqua is blown skyward, buffeted by convection forces and atmospheric winds that are constantly blowing across the planet. Eventually, Aqua finds itself streaming east from the islands, 10 kilometres above the ocean, moving along a ribbon of moisture that is like a great atmospheric river.

Reaching the coast of North America, Aqua moves inland until it encounters the upthrust of the Rocky Mountains. The cloud Aqua is in begins to cool, condense and finally liquefy, and the water molecule falls toward the land as part of a drop of rain. On striking Earth, Aqua slithers into the soil, pulled by the forces of gravity, moving erratically around grains of sand that loom like miniature planets.

As Aqua sinks into the soil, it encounters a slender rootlet of a tree, which slurps Aqua up into its xylem tissue, drawing the molecule by capillary action up through the trunk into the branches. Eventually Aqua ends up in one of the seeds in a pine cone. A bird pecks at the cone, dislodging and swallowing the seed containing Aqua. As the bird flies south on its annual migration, it absorbs Aqua into its bloodstream.

Resting in a tropical rainforest in Central America, the bird is preyed upon by a mosquito. Aqua is sucked into the mosquito's gut, and as the blood-laden insect drops close to a creek, it is snapped up by a sharp-eyed fish, which incorporates Aqua into its muscle tissue. An Indigenous fisher spears the fish and triumphantly carries it, and Aqua, home for a meal. And so it goes, the endless, eventful peregrination of every molecule of water. □

This cycle between air, water and forest is most striking in the tropical rainforest of the Amazon basin. This largest tract of tropical forest on Earth is an important heat engine, cycling massive amounts of water each day through the forest as trees and other plants absorb water and then breathe it out. Every year, a single large tree can move 700 tonnes of water out of the soil, plumbed through its trunk into the forest canopy and, ultimately, the atmosphere. There is so much moisture locked in the Amazon's forests that it acts like a green ocean, seeding the clouds sailing above it with moisture, just as clouds form above the seas. The Amazon forest's influence is far-reaching, fuelling major atmospheric and oceanic currents, which in turn drive world climate and weather patterns.

The Circulatory System of Earth

Across continents, the network of waterways resembles the circulatory system of a body. And in fact, that is the role that lake and river systems perform. Water that runs off after rain or from melting snow or from plant roots accumulates in ditches and creeks, which carry it to the rivers draining into lakes or the oceans, where it evaporates back into the atmosphere. Patterns of rootlets, roots and branches; rivulets, creeks and rivers; veins and capillaries in living tissue—they all reflect the same physical realities and bind us all together in the Earth's vital processes. In the words of aquatic ecologist Jack Vallentyne:

> If water is the blood of Mother Earth and soil the placenta, river courses are veins, oceans are compartments of the heart and the atmosphere is a giant aorta. Comparing Earth beats to human heart beats, the life expectancies of rivers would range from millions to billions depending on whether Earth beats were measured in days or years.

Water sits still as well as flowing—wrapping a film around minute particles of soil, pooling in the interstices of rocks and gathering to a greatness in deep underground aquifers, which have existed since the dinosaurs roamed the land. This "fossil" water may move a few metres every thousand years; the water in the aquifer below the city of London is deemed to be twenty thousand years old. Water isn't constantly generated de novo; what is here on Earth has always been here. But the transformational process by which it maintains life, cartwheeling around the planet, moving from cloud to rain to ground and back again in the hydrologic cycle, has not always existed. It is the product of a multitude of factors—temperature, chemistry, soil and life itself.

The First Flood

In the early life of Earth, the atmosphere was too hot for water to exist in liquid form. Water vented from volcanoes was vaporized, and only after tens of millions of years, when the atmosphere had cooled enough, was water able to condense into clouds. Eventually, those clouds were able to release their contents by raining onto the rock that formed the surface of the planet.

Imagine the dry, lifeless rockscape that was then Earth: immense mountains pierce the sky; deep trenches scar the surface. As the rain falls relentlessly, water accumulates in every depression, filling each one and flowing down toward the next containment. Pulled by gravity, water overflows the depressions, becoming creeks and rivers, dragging rocks along, scouring out channels, always running down toward lower places.

After millions of years, fresh water covered most of the Earth. The relentless flow of water dissolved compounds out of rock and wore away minute quantities of elements, washing them away to the largest bodies of water. The salty seas were formed by this imperceptible accretion, an enormous change achieved by infinitesimal alterations over immense periods of time.

Life began in the Archean period, 3.8 to 2.5 billion years ago, and even in these early days life seems to have played a part in maintaining Earth's supply of water. During that period oxides in basalt rock continually reacted with carbon dioxide and water, producing various carbonates (compounds of oxygen and carbon) of sodium, potassium, calcium, magnesium and iron and releasing hydrogen into the atmosphere. Since hydrogen is extremely light and cannot be held by the gravitational pull of the planet, it was lost into space. Had this reaction continued for a billion years or more, all of the planet's water might have been lost, and Earth's atmosphere would be like that of Mars. Instead, cyanobacteria (and, later, algae and plants) capitalized on the abundant resources available to them—water, sunlight and carbon dioxide—to fuel a chemical reaction that would change their world. Photosynthesis produced oxygen as a by-product, but it also held some of the hydrogen from water in the carbon ring of glucose, thus clutching hydrogen to the planet. In addition, the free hydrogen produced by the oxidation of iron in rock was exploited by bacteria as a source of energy. Oxygen, hydrogen and sulphur react chemically to produce water and hydrogen sulphide, which has recoverable energy within its structure. Thus, the forces of life may very well have prevented the desiccation of the planet by capturing the hydrogen that is necessary for water and thereby preventing it from drifting into space.

Our Need for Water

Life is animated water.

—VLADIMIR VERNADSKY, quoted in M.I. Budyko, S.F. Lemeshko and V.G. Yanuta, *The Evolution of the Biosphere*

Like air, water is essential to our survival. But whereas the lack of air will kill us within minutes, water takes longer to make its necessity known to us. Deprived of water for a few hours, especially after exercise or on a hot day, we notice our throats becoming parched; our bodies are urging us to take a drink. If no water is available, we might survive for as long as ten days, depending on the ambient temperature, our degree of activity and the clothes we are wearing. But in the end we would die a terrible death. Water is the elixir of life; without it, this planet would have remained barren.

Living beings need this elixir because they are made of it. Protoplasm, the living matter of all plant and animal cells, is mostly water. The average human being is roughly 60 per cent water by weight, nearly 40 litres of it carried in trillions of cells. Three-fifths of the water in our bodies is inside cells and is called intracellular fluid, and two-fifths is outside cells, in blood plasma, cerebral spinal fluid, the intestinal tract and so on. The proportion of water by weight varies with age and gender as shown in Table 3.1.

The reason for the difference lies in the proportion of fluid that is inside and outside of cells and the proportion of body fat, for fat cells contain less water than other cells.

||||||||||||||||||||||||||||||||||||||||||||||||||||||||||||||||||||||||||||||||||||||||||||||||||||

Table 3.1: Proportion of Weight in Water for Humans Grouped by Age and Gender

GROUP	% OF WEIGHT IN WATER
Babies	75%
Young men	64%
Young women	53%
Elderly men	53%
Elderly women	46%

Though we appear to be solid, we are in fact liquid bodies, similar in a way to gelatin, which also seems to be solid but is in fact largely water, "gelled" by the presence of an organic material.
    —DANIEL HILLEL, *Out of the Earth*

Basically, each of us is a blob of water with enough macro-molecular thickening to give us some solidity and to keep us from dribbling away. Every day, about 3 per cent of the water in our bodies is replenished with new molecules. The water molecules that perfuse every part of our bodies have come from all the oceans of the world, evaporated from prairie grasslands and the canopies of all the world's great rainforests. Like air, water physically links us to Earth and to all other forms of life.

Preserving the Balance

Although we live on land, we are creatures of water, and, as such, we have our own hydrologic cycle. We lose moisture with every breath, in each bead of sweat, in every tear we shed, every time we urinate and defecate. We can replenish some of this water ourselves through metabolic processes such as the breaking down of carbohydrates and fats, which produces carbon dioxide, water and energy. But metabolic production of water only accounts for 11.5 per cent of the normal requirement of 2.5 litres a day. The rest usually comes from the fluids we drink (52.2 per cent of our daily needs) and from solid food (36.3 per cent). This intake is balanced with our daily losses—about 1.5 litres in urine, 0.9 litres in expired air and sweat, and 0.1 litres in feces.

Our bodies are perpetually on water alert, because our daily intake must be exquisitely matched to our daily output. When you start to become dehydrated, the concentration of salts in your body fluids begins to rise. A small change is enough to induce the posterior lobe of the pituitary gland to release the hormone adiuretin (ADH, or vasopressin). ADH acts directly on the kidney, inducing it to decrease the excretion of water.

Other biological alarms are set off when dehydration reduces the volume of the blood. Stretch receptors monitor blood volume inside the heart and send signals to the "thirst centre" of the hypothalamus in the brain to inhibit the production of saliva. Dryness in the mouth registers in our consciousness as "thirst," stimulating us to drink. The cottony mouth sensation is one of the early signs of dehydration, which commonly results after extensive bleeding, burns or diarrhea, along with profuse sweating.

If you drink too much water, these alarm systems work in reverse. When the concentration of salts in your body fluids becomes diluted, the production of ADH is inhibited, stimulating the kidneys to excrete more water. A more dilute concentration of urine is sent to the bladder, and usually the excess is eliminated within an hour.

As well as maintaining the water balance in the body, the kidney plays a major housekeeping role in purifying the crucial fluids in blood. It removes dissolved metabolic wastes, such as ammonia from the breakdown of amino acids, urea produced in the liver from degraded protein products, uric acid from nucleic acid, and phosphoric and sulphuric acids from protein by-products. These toxic compounds are filtered out of the blood and flushed away. About 1.2 litres of blood passes through the kidneys every minute for a total of almost 2000 litres per day. The nephrons and their network of minute filtration structures filter 180 litres of blood each day.

As you will see in chapter 5, on energy, one property of water, its high absorption of heat to change from a liquid to a gaseous state, plays a critical role in regulating body temperature. Water within the body reaches the surface of the skin by diffusion (*perspiratio insensibilis*) or via sweat glands that are activated by the

autonomic nervous system, which functions without our awareness. After it reaches the surfaces of the skin as sweat, the water
evaporates. Evaporation requires energy, so drying sweat takes
heat from the body as energy, thereby cooling the skin. Evaporation of a litre of fluid requires 2428 kilojoules of heat.

There is a remarkable equilibrium between your body and its
surroundings. The inside and the outside of your body combine
to manage the ebb and flow of water within and around you.
Ambient humidity and air temperature, together with your level
of physical activity, determine how much water moves through
your skin into the surrounding air. In the same way, external and
internal conditions regulate the water you imbibe and the water
you eliminate. The same is true of all other creatures; this lifelong balancing act is part of a global circus, a performance stage-
managed by the planet and its inhabitants together.

Water enters our bodies, circulates through it to the rhythm
of the heart, ceaselessly carrying food, fuel, and cellular and
molecular detritus to and from various organs of the body. Water
seeps through our skin, escapes from our lungs as vapour and
exits every opening in the body. It then re-enters the hydrologic
cycle, trickling into the soil, entering plants, evaporating into
the atmosphere, entering bodies of water. In this way, water circulates endlessly from the heavens to the oceans and land, held
briefly within all living things before continuing the cycle. You
might see the whole enterprise of life as just a vehicle for the
transformation of water. If a hen is the egg's way of being born,
then human beings are the way water molecules get to talk to one
another.

The Special Properties of Water

Looked at closely, water molecules turn out to be very strange
things. Water is so familiar that most of us accept its behaviour as
normal. But to a physicist water is an anomaly. For example, water

is liquid at room temperatures. That is quite odd; a compound such as hydrogen sulphide, which has a low molecular weight similar to that of water, becomes a gas at $-60.7°C$. Whereas most substances contract when they solidify, water expands. This means water is less dense as a solid than as a liquid. Water has other unusual properties, such as high melting, boiling and vaporization points.

Life depends in toto on water's constancy. The ability of water to ab-sorb large amounts of energy buffers photosynthesis in cytoplasm and the transfer of oxygen in animal blood from chaotic flux; moderates the Earth's climate by using oceans and lakes for heat storage; eases seasonal change and our bodies' adaptation to it by slowing, without shocks, the change of weather; and protects plants like cacti from boiling under desert skies. Most of all, water's specific heat, heat of vaporization and heat of fusion give life its ability to maintain in hard times. Without these molecular traits, climatic extremes would turn living creatures over to their Maker at unprecedented rates.

—PETER WARSHALL, "The Morality of Molecular Water"

Water has one property that is particularly striking: the amount of heat required to raise the temperature of a unit of water by $1°C$ is ten times higher than for iron, thirty times higher than for mercury and five times higher than for soil. This prop-erty makes water an effective "sink" for heat; it absorbs large quantities of heat and then radiates it out. Because of this prop-erty, large bodies of water, such as lakes and oceans, absorb a great deal of heat in summer, release it in winter and thus modu-late surface temperatures. Ocean currents absorb large amounts of heat in the tropics and transport it to temperate regions, where it warms the surrounding air. When the water reaches the polar areas, it is cooled and then moves back toward the equator, where it will lower air temperatures as it absorbs more heat. The planet's water supplies have other effects on climate as well: the whiteness of snow and clouds reflects sunlight away from the Earth, whereas water vapour behaves as a greenhouse gas and reflects heat back onto the surface.

Water's special qualities are the result of the strong attraction between molecules of water, giving it a high internal cohesion. At the molecular level, water is a deceptively simple substance, made up of two atoms of hydrogen combined with a single atom of oxygen. The hydrogen atoms do not line up with the oxygen atom in a linear array as H-O-H; instead, they form a 105-degree angle to each other (Figure 3.2). They are on one side of the molecule and have a positive charge, while the large oxygen atom bulges out at the other end and has a negative charge. Thus, the molecule is said

Follow the Water

Water is so fundamental to life that the search for extraterrestrial life is essentially a search for water. Guided by the mantra "Follow the water," researchers are on the lookout for water throughout our solar system. Thus far, the leading candidate is Mars, with its polar ice caps and evidence of the largest-known flood in the solar system. Three and a half billion years ago, Mars was awash, but where did the water of this ancient flood go? In 2002, the *Mars Odyssey* may have found the answer when it discovered enough ice-studded soil to fill Lake Michigan twice over just below the planet's surface. Tantalizing new images from Mars rovers and spacecraft suggest that there may also be underground reserves of water that break through the planet's surface as springs.

Mars is 700 billion billion tons of iron and rock, wrapped in an unfamiliar landscape of canyons, craters and calderas. Nonetheless, the most compelling thing we could find on this enormous, to be dipolar, like a tiny magnet. This dipolar arrangement gives the hydrogen atoms an attraction for an oxygen nucleus on another water molecule, a kind of chemical affinity that is called hydrogen bonding (Figure 3.3). The extraordinary ramifications of the dipolarity of water can be seen in the formation of crystals or snowflakes. There are so many possible ways this simple molecule can combine that the shapes of snowflakes appear to be infinite. It took a curious, self-educated young farmer to show us this was so. In 1885, Wilson "Snowflake" Bentley became the first to photograph a single snow crystal. He went on to photograph over five thousand snowflakes and, indeed, never found two alike. His images are still among the best ever taken of water in its crystalline state.

orange orb would be a microgram of wet chemistry able to reproduce, move, grow, and evolve.
—SETH SHOSTAK, astronomer, SETI Institute

Although Mars is the front-runner as a potential home of extra-terrestrial life, other planets and satellites are in the race, too. Europa, one of Jupiter's moons, is not only ice-covered but also thought to have an ocean about 20 kilometres below its surface. If, as is surmised, volcanic vents below this sea are spewing hot, nutrient-rich water into its abyss, the recipe for life (water, plus an energy source and nutrients) might just exist. The Jovian moons Ganymede and Callisto may also house great oceans under their icy crusts, but recently all eyes have turned to Enceladus, one of Saturn's moons. Images captured by the spacecraft *Cassini* show tantalizing images of spouting geysers, making Enceladus the first world other than Earth with the most convincing evidence of liquid water. □

||||||||||||||||||||||||||||||||||||||||||||||||||||||||||||||||||||||||||||||||||||||||||||||||||||||||||||

Figure 3.2: The atomic structure of a water molecule.

Slight negative charge at this end

Slight positive charge at this end

||||||||||||||||||||||||||||||||||||||||||||||||||||||||||||||||||||||||||||||||||||||||||||||||||||||||||||

Figure 3.3: The formation of hydrogen bonds between water molecules.

Figures 3.2 and 3.3 adapted from Cecie Starr and Ralph Taggart, *Biology: The Unity and Diversity of Life,* 6th ed. (Belmont, CA: Wadsworth, 1992), p. 27.

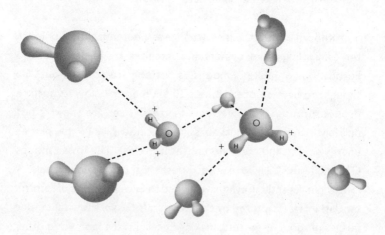

Water molecules cling to each other, but unlike actual chemical bonds, the hydrogen bonds change constantly. A molecule may change hydrogen-bonded partners 10 billion to 100 billion times a second, thus linking adjacent molecules in a fleeting embrace. These rapid-fire interactions are like a frenetic dance of water molecules. Chemist Richard Saykally describes the dance in this way:

> ... it's like water has two hands and two feet. The hands of water are the hydrogens that are more or less positively charged, and the feet are the electron pairs that are the negative part associated with oxygen. And these two hands want to grab the feet of two other water molecules, and the two feet want to interact with the hands of two other water molecules. So in each water molecule, hydrogen bonds to four others, making very extensive networks in the liquid.

This constant shifting makes liquid water so stable that it requires a great deal of heat energy to enable molecules to break free as a gas.

In ice, each molecule of water grabs the "hands" and "feet" of its four nearest neighbours. This forms a tetrahedron, a pyramid with four triangular faces. What happens to this shape when the ice melts? According to Richard Saykally and other chemists, liquid water looks surprisingly similar to ice. The only difference is that about 10 per cent of the hydrogen bonds are broken. The molecules with broken bonds continue to reform, break and move around. The ability of water to retain most of the hydrogen bonds may be the secret to explaining why water, liquid water in particular, has so many unusual properties. When water freezes, the crystal it forms has more space between molecules than does the liquid phase. Thus, ice expands and therefore floats rather

||||||||||||||||||||||||||||||||||||||||||||||||||||||||||||||||||||||||||||||||||||||||||||||||||||||||||

Figure 3.4: The basis for water's ability to dissolve salt.

Adapted from Cecie Starr and Ralph Taggart, *Biology: The Unity and Diversity of Life*, 6th ed. (Belmont, CA: Wadsworth, 1992), p. 29.

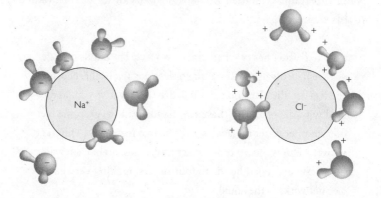

than sinking, as most freezing liquids do. That's why instead of forming along the bottom of lakes and rivers, ice rises to the top. More important, the ice that forms insulates the rest of the body of water and keeps it in liquid form, enabling aquatic life to survive the winter.

Water is a universal solvent, dissolving many minerals and organic compounds. It does so because the dipolarity of water molecules enables them to surround atoms or molecules at sites of electric charge (Figure 3.4). Because it is a universal solvent, water is an effective agent in weathering and decomposing rocks. As it percolates through the soil, it dissolves nutrients and materials and carries them with it. It also makes cellular molecules soluble and thereby transports materials within living organisms. But water is more than a solvent; it also enters into metabolic reactions to become part of breakdown products or is released as a by-product when large molecules such as fats are broken down.

Many of the systems for quantifying and measuring the physical world use water as their reference point, acknowledgement of its unique place on this planet and in our lives. The metric system of weights and measures defines 1 gram as the weight of 1 millilitre of water. The centigrade scale sets 0°C as the freezing point of water and 100°C as its boiling point. Units of energy are measured in calories, defined as the amount of energy needed to raise the temperature of 1 cubic centimetre of water by 1°C; in food, 1000 calories or 1 kilocalorie is 1 Calorie.

Freshwater Supplies

Water, water, everywhere,
Nor any drop to drink.
  —SAMUEL TAYLOR COLERIDGE, *The Rime of the Ancient Mariner*

Human beings, like most other terrestrial animals and plants, have an absolute need for fresh water—and that is the rarest form of water on Earth. More than 97 per cent of the planet's water is salty, toxic for terrestrial organisms, which require sweet water to sustain life. Of the water that is sufficiently free of salt to drink, more than 90 per cent is locked away in glaciers and ice sheets or is deep underground. Only about 0.0001 per cent of fresh water is readily accessible.

Human beings lived along waterways that they used for food and travel long before there was a history. We can infer this fact from prehistoric middens and sites of habitation. And it was on the great flood plains that humans first established permanent settlements, exploiting for agricultural use the regular floods

||||||||||||||||||||||||||||||||||||||||||||||||||||||||||||||||||||||||||||||||||||||||||||||||||||

Table 3.2: Distribution of Water on Earth

LOCATION	VOLUME (KM³)	% OF TOTAL
Oceans	1,322,000,000	97.2
Ice caps and glaciers	29,200,000	2.15
Groundwater (below water table)	8,400,000	0.62
Freshwater lakes	125,000	0.009
Saline lakes and inland seas	104,000	0.008
Moisture in soil (above water table)	67,000	0.005
Atmosphere	13,000	0.001
Stream channels	1,250	0.0001
Total liquid water in land areas	8,630,000	0.635
World total (rounded off)	*1,360,000,000*	*100.00*

that fertilized the deltas. At the junction of the Tigris and Euphrates Rivers in Mesopotamia the first civilizations arose, followed by settlements along the Nile River. Other great rivers of Earth, such as the Amazon, Mississippi and Ganges, have provided a living to Indigenous people for millennia. The origins of villages and towns are tightly linked to the presence of water; even today most major cities are next to oceans, lakes or rivers. Elsewhere people have had to learn how to find water, digging wells, catching and storing rainwater, even trapping mists and clouds so that they can grow crops in arid regions.

A river rose in Eden to water the garden.
　—GENESIS 2:10

It is the hydrologic cycle that allows the unpotable water of the oceans to rise into the skies as sweet water and sustain life on the land. Even though a minute quantity of potable water is readily accessible to land organisms, the hydrologic system draws fresh water from the oceans and land and returns it as rain and snow. Each year, over 113,000 billion cubic metres of water fall to Earth, enough to cover all the continents to a depth of 80 centimetres. Two-thirds of this amount evaporates back into the atmosphere, while surface and subsurface waters are replenished by the rest. That water is not evenly distributed, of course; some regions get a great deal of water and others do not get much at all.

The amount of water determines the nature and abundance of vegetation in each region. The vast continent of Australia, for example, is often referred to as "underpopulated"; in fact, it is too poor in water in relation to its land base to support a larger human population. In contrast to the great river systems meandering through North America—the Mississippi, the Columbia, the Mackenzie—an enormous desert occupies the centre of Australia.

Very great rivers flow underground.
　—LEONARDO DA VINCI

Canada is one of the "have" nations of the world, with more than half of the planet's fresh water by area and 15 to 20 per cent by volume. It may be hard to believe, but this country was blessed by the last ice age, more than eight thousand to ten thousand years ago, when glaciers gouged out the land to create depressions into which water settled. The Great Lakes alone, which Canada shares with the United States, contain nearly 20 per cent of all the surface fresh water on Earth and serve the needs of some 40 million people living around them. Canadians have a volume of 130,000 cubic metres of flowing river water a year per person, compared with 90 cubic metres a year per person in Egypt.

Where Alph, the sacred river, ran
Through caverns measureless to man
Down to a sunless sea.
  —SAMUEL TAYLOR COLERIDGE, "Kubla Khan"

Water defies human boundaries and human ownership. Sweeping invisibly through the air as vapour, flowing across the surface of the planet, percolating through soil, seeping into underground caverns and channels, it moves in its own mysterious ways. The mobility of water complicates human affairs at many levels. Neighbours tapping into the same aquifer for well water must share a source that has little or no relationship to their property lines. Factory waste entering a stream or draining into the soil at a specific point has ramifications for plants, animals and human beings over a large, often unpredictable area. One of the greatest bodies of fresh water in the world, the Great Lakes, along with their rivers, are administered by two federal governments and two provincial and four state jurisdictions, while dozens of cities and towns have a vital stake in the water. The Nile River in Africa flows through seven countries, each of which draws on it for irrigation and drinking water while flushing sewage and effluent into it. Egypt, the last nation downstream, inherits the collective consequences. Not surprisingly, water, not oil, is the real flashpoint issue over which wars will be fought in the Middle East and elsewhere.

The Oceans

All the rivers run into the sea, yet the sea is not full.
—ECCLESIASTES 1:7

Together with the sun, the oceans drive the planet's climate. Whereas the temperature of the air changes rapidly, the oceans absorb massive amounts of energy and release it slowly. Thus, the oceans stabilize the temperature of Earth. In the mid-latitudes, huge wind-driven gyres (circular systems of currents) transport heat polewards from near the equator, ameliorating terrestrial

Figure 3.5: Ocean currents of the mid-latitudes.

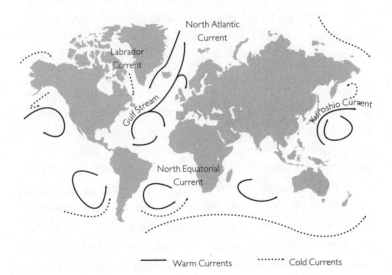

temperatures and weather. The warm Kuroshio Current flowing from the western Pacific Ocean south of Japan across to North America affects weather as far inland as the Midwest and from California to Alaska. Its counterpart in the Atlantic, the Gulf Stream, meanders north from the Gulf of Mexico, bringing the gift of warmth to Canada's east coast. South of Newfoundland the Gulf Stream meets the cold Arctic waters of the Labrador Current; the meeting of warm and cold currents creates the famous fogs of this region. Moving across the North Atlantic, the Gulf Stream divides; the northern branch wraps the British Isles, bringing palm trees to sheltered spots in northwest Scotland and a milder climate to the whole country than that latitude would otherwise enjoy. The southern arm curves south past Portugal to join the Northern Equatorial Current (Figure 3.5).

There are deepwater currents as well—great aquatic conveyor belts moving waters of differing temperatures and salinities. In winter in Antarctica for example, as sea water freezes into ice, salt is concentrated and so water becomes supercooled and superdense, sinks to the bottom and flows in immense, slow-moving "rivers" carrying ice-cold water around the Indian Ocean, past the tip of Africa northward deep in the Pacific trench.

Sweeping over great distances, the currents carry eggs and larvae of animals that have evolved to ride with the movement of the ocean. The currents waft carcasses of plants and animals, as well as minerals and elements and soil, toward the ocean floor. Humans have long taken advantage of the ocean currents: accepting gifts delivered to our coastal doorsteps, fishing in the nutrient-rich zones where warm and cold currents meet, and using these currents as moving highways, routes to profitable trade. But they are far more than that; when we use the currents, we are in touch with the great forces of the planet—its rotation in space, the prevailing winds, the slow, curling drift of ocean water transporting heat, maintaining the planet's atmospheric equilibrium. Connecting continent to continent, pole to pole, the currents are like a living web, moving and winding and mixing, wrapping itself perpetually around the whole world.

Water Works

When we look at the wonderful array of plants and animals on Earth, the overwhelming lesson is that life is opportunistic, taking advantage of niches through mutation and new combinations of genes. Plants and animals have evolved to exploit both marine and freshwater environments. The oceans are filled with plants—immense kelp forests and massive blooms of phytoplankton that are the base of the marine food chain. The abundance of forms that co-operate to make the coral reef community, the forests of mangroves lining the ocean beaches, the gatherings of creatures in estuaries—all attest to the power of evolution to hone organisms for diverse habitats. On land, plants and animals alike have found strategies to flourish where water is abundant and where it is rare. Species are found in the ice of polar sheets, on arid mountaintops and in the dry heart of the desert. Anadromous fishes such as eels and salmon have evolved life cycles that exploit both marine and freshwater environments, and numerous species inhabit both water and air or water and land at different stages of their life cycles. Many species have adaptations to help them retain each precious molecule of water. Instead of broad leaves, cacti have needles with a small surface area and a tough outer coating to minimize water loss; the extremely hairy nostrils of camels trap and condense moisture as it is breathed out. The microscopic tardigrade, or water bear, can live for decades, possibly even centuries, in an inactive, desiccated state. Within hours of receiving the gift of even a single drop of water, the tardigrade revives and resumes its normal state. The eggs of brine shrimp, *Artemia*, can be dried and still remain viable for years. When introduced to water, the eggs hatch and thrive. (Thus the secret of how colonies of "sea monkeys" could be shipped to eager comic book readers via the mail.) Many organisms can cope with a scarcity of water, but no species has evolved to do without it, and no species has been as imaginative and as demanding in its use of water as human beings.

Use and Misuse

Every day each of us requires a certain amount of water to compensate for what is lost and to maintain a constant internal balance, but that amount is a small fraction of the water that we use for other reasons or that is used on our behalf. Much more water is used in rich countries than in poor countries, however. To further the inequity, much of the water in developing countries is of poor quality. As of 2020, it's estimated that 2 billion people live without clean drinking water and that 829,000 people, including 297,000 children, die every year from water-borne diseases. That's about 800 children a day, or one classroom of children every hour.

Many water-rich countries, such as Canada, use water as if it were limitless. We often meet our food, energy and material needs through the copious use of water, whether we know it or not. On any North American dinner table irrigation may have produced the vegetables, hydroelectric power may have cooked them, and the dish they're served in may have taken litres of water to manufacture. And the amounts of water used for agriculture and manufacturing can be staggering. For example, it takes about 20,000 litres of water to produce 1 kilogram of coffee, 11,000 litres for a quarter-pounder hamburger, between 2000 and 5000 litres for 1 kilogram of rice, and 2000 to 4000 litres for 1 litre of milk. We grow twice as much food today as we did a generation ago, but we use three times as much water to grow it. To quench the thirst of these crops, many are highly irrigated, and much of the water is being pumped from underground reserves, which take eons to replenish. Industry uses

Message in a Bottle

The sale of bottled water has soared in recent years—in 2019, Americans spent a staggering $19 billion for bottled water, which costs on average a thousand times more than tap water. Savvy marketing has positioned brands of water as safer, healthier or trendier. In reality, bottled water is a marketing phenomenon and an environmental disaster. Despite the promise of a healthier alternative, there is no guarantee that bottled water is safer than tap water. Regulation of the industry is uneven, and tests have found bottled water with coliform bacteria, arsenic and synthetic chemicals. Testing for tap water is much more rigorous than it is for bottled water.

The source of water is also a problem. About one-quarter of bottled waters are simply bottled tap water, while other "manufacturers" pump their water from springs and aquifers, critical reserves of water for our planet.

And then there are the plastic bottles. Each of those bottles takes energy and resources to manufacture, transport and recycle or dispose of. In the United States alone, it takes 1.5 million barrels of oil for a year's supply of bottles, and nine out of ten of these bottles end up as garbage or litter. The plastic may also contaminate the water it is meant to hold. A study by *Consumer Reports* showed that eight of the ten plastic water jugs tested leached the endocrine disrupter bisphenol A into the water.

When so many of the world's people lack the basic human right of safe drinking water, it seems misguided to direct so much money and so many resources into designer bottled water. Surely it would be better to direct this energy into advocating for tap water that is safe, accessible and affordable to all. □

water on our behalf in a multitude of ways—as part of the reaction mix where chemicals are used or as a medium for carrying material such as wood fibres in pulp or for washing away excess material.

The fate of the Great Lakes illustrates the dilemma we face. For the Indigenous people who lived on the shores of the Great Lakes, the waters were sacred, an endless source of food and water and a great waterway to other parts of the continent. The arrival of Europeans began a different relationship with the lakes. As forests surrounding the lakes were cleared, watersheds were altered and the quality of the water was diminished. Once-abundant native fish were reduced by overfishing, and new kinds of fish were introduced to replace them. When the Welland Canal was built to enable boats to bypass Niagara Falls, parasitic lamprey were able to hitch a ride into the upper lakes, where they devastated fish populations. More recently, exotic zebra mussels have grown with explosive force since being introduced from water ballasts taken on by ships in other parts of the world. As a result, the Great Lakes are now in a period of massive upheaval as alien species alter the ecological makeup of the lakes. The waters have been commandeered for agricultural irrigation, for industry and for drinking water; at the same time they have become a repository for sewage and effluent from surrounding urban populations. Shorelines have been altered by urban development, exposing them to rapid erosion, while the lakeside marshes that once filtered out organic matter and fed native wildlife have been filled in, paved over or polluted. The relentless increase in population has added an intolerable burden to the lakes' ability to support healthy life forms.

Lake Erie, around which about 11.6 million people live, has been the focus of scrutiny and scientific study for decades. Erie has the greatest stress of all of the Great Lakes, with seventeen large cities on its shore and an enormous impact from urbanization, industry and agriculture. It reached perhaps its lowest point in June 1969 when the Cuyahoga River, which flows into the lake, caught fire. The media declared the lake dead, and political pronouncements to clean up Erie and other Great Lakes followed. And there have been improvements over the years, with stricter regulations on effluents, better sewage control and treatment, reduction of phosphates in detergents and the banning of pesticides such as DDT. Still, serious problems persist.

Of great concern are the "hand-me-down" chemicals such as PCBS or PBDES (found in flame retardants) that are extremely stable and are passed on through food chains. For years, biologists still could not determine why they were finding so many organisms with unusual deformities, reduced size, reproductive failure or aberrant parental behaviours, such as birds being inattentive to their eggs or chicks. Studies now show that many of the persistent chemicals in the Great Lake food chains affect the endocrine system, which regulates hormones in glands related to metabolism and reproduction. This disruption interferes with the sexual development of wildlife. For instance, salmon in the Great Lakes have greatly enlarged thyroid glands, which result from inadequate levels of thyroid hormones. This deficiency, in turn, disrupts reproduction and the normal development of eggs and offspring. Evidence points strongly toward persistent chemicals in the food chain that block proper functioning of thyroid hormones.

Humans, of course, are not immune from these persistent toxins. Studies of women living near Lake Michigan showed that the higher a mother's consumption of Lake Michigan fish, the lower her baby's birth weight.

In Toronto, which depends for its drinking water on Lake Ontario, the last in the chain of five lakes, many Torontonians now pay for bottled water rather than drink water from their taps.

If we lack the knowledge to keep water pure, then it makes sense to control those factors that we *know* cause problems with water and to protect nature, which has provided clean water since the beginning of time. Water is integral to supporting and maintaining life on this planet as it moderates the climate, creates growth and shapes the living substance of all of Earth's creatures. It is the tide of life itself, the sacred source.

We are water—the oceans flow through our veins, and our cells are inflated by water, our metabolic reactions mediated in aqueous solution. Like amphibians and reptiles, we mammals have moved from continuous immersion in water but cannot avoid the need for water in reproduction. In the most intimate of human acts, spermatozoa are set free in seminal fluid to swim toward their target, the fertilized egg embeds itself in the rich, blood-lined walls of the uterus, and the growing embryo floats in a primeval sea of amniotic fluid, sprouting gills in a recapitulation of our aquatic origins. Water is created in the metabolism of life; we absorb it from solid food and from any liquid we imbibe. As air is a sacred gas, so is water a sacred liquid that links us to all the oceans of the world and ties us back in time to the very birthplace of all life.

Of all our natural resources, water has become the most precious... In an age when man has forgotten his origins and is blind even to his most essential needs for survival, water along with other resources has become the victim of his indifference.

—RACHEL CARSON, *Silent Spring*

The Loss of Life in Lake Erie

Within our lifetimes, the lakes have changed with dramatic speed. In the late 1940s, I lived in the centre of Canada on the shores of Lake Erie in Leamington, a town near Point Pelee, the southernmost part of the country. Each spring, an immense hatch of mayflies emerged from the lake, filling the air with the throb of their wingbeats and engulfing homes and roads. Their carcasses piled up on the shores to a depth of a metre and more. Fish in the lake went into a frenzy of feeding, while birds, small mammals and other insects feasted on this annual banquet. In a decade, this enormous biomass that gave tangible evidence of the water's productiveness was gone and the lake was declared "dead." Eutrophication—excess algal growth stimulated by phosphates—had choked off oxygen from other aquatic organisms, while DDT from farm runoff polished off invertebrates.

In the late 1950s, while crossing the Niagara River on a train, I glanced down into the gorge to see fishers pulling fish from the water as fast as they could cast a hook. They were intercepting the annual spawning run of silver bass, a wondrous sight of water jammed with the flashing bodies. Again, by the 1960s, the bass were gone, casualties of overfishing and pollution. Today the Great Lakes are reeling under the impact of introduced fish and plant species, agricultural and industrial effluents, and development of the watersheds that recharge them. □

4

Made
From the Soil

With the sweat of thy brow shalt thou eat bread
till thou return to the earth, for out of it wast thou taken,
for soil thou art and unto soil shalt thou return.
—GENESIS 3:19

EARTH IS BOTH the planet we live on and the material we live
from. In our origin stories it is the stuff of our existence:

> God Jahweh formed man out of the soil of the earth and blew
> into his nostrils the breath of life, and man became a living
> soul. And God Yahweh planted a garden in Eden in the east and
> placed the man therein... God Yahweh took the man and put
> him in the Garden of Eden to serve and preserve it.

The first man of the Bible is named Adam, from the Hebrew
adama, meaning "earth," or "soil." The first woman, created from
Adam's rib, is Eve, from *hava*, meaning "living." Together they
make the eternal connections: life comes from the soil; the soil is
alive. In other creation stories the fundamental connection of soil
and life is expressed in different ways; in some, the first human
is fashioned from material produced by earth—carved from

wood, moulded out of cornmeal, shaped from seeds, pollen and sap. Sometimes the turtle or the water beetle carries soil up from the bottom of the sea; sometimes humans emerge from the underground womb of the Earth, as rain and sun, sand and seed combine into the shape of our first ancestor. One way or another, we are Earthenware. These stories tell us the truth: the soil is the source of life. Throughout the ages it has been treated as precious, even sacred, because of the gifts it gives us. According to Daniel Hillel:

> Worship of the Earth long predated agriculture and continued after its advent. The Earth was held sacred as the embodiment of a great spirit, the creative power of the universe, manifest in all phenomena of nature. The Earth spirit was believed to give shape to the features of the landscape and to regulate the seasons, the cycles of fertility, and the lives of the animals and humans. Rocks, trees, mountains, springs and caves were recognized as the receptacles for this spirit.

But some children grow up to despise their mothers; moving out in hope of bettering themselves, they try to conceal their origins, even from themselves. As nations become more industrialized, people see earth as "dirt," a filthy material that "soils" them. In our urban habitat, dominated by concrete and asphalt or carefully manicured lawns, we find ourselves separated from the source of life. Accustomed to thinking of food as a packaged commodity supplied by supermarkets, we also forget that all of our food comes from the earth. Uncoupled from the earth, we forget a fundamental truth: every bit of the nutrition that keeps us alive was once itself alive, and all terrestrially supplied nourishment comes directly or indirectly from the soil. As botanist Martha Crouch points out, our relationship with food is the most intimate of all the connections we have with other beings, for we take it into our mouths and actually incorporate it into our cells. Every part of our bodies, as well as the sugars, fats and enzymes that drive the metabolism of our cells and fuel life, are constructed out of building blocks absorbed from the carcasses of other life forms. Deprived of other beings to eat, we begin to starve and thus to consume ourselves; if starvation continues, we will die within seventy days.

Earth, soil, dirt, ground, land: these terms embrace ideas of extraordinary complexity. Hidden within them is our sense of our origins, our place, our dependence on the soil beneath our feet; when we unearth the truth, we find pay dirt. In medieval times dirt meant excrement, manure or fertilizer; a good farmer tended the land by spreading the dirt, or soiling it, as the word was used then. What we call dirt today, in the wisdom of our tongue's history, is the fertile ground, the tended source of food, the planet's life. It is a support system, a supply network, like the other elements on which we depend, and most peoples who have ever lived on Earth have acknowledged that truth and lived by it. Ground is something solid, the space we gain or lose in battle, the place where we take a stand, the basis for all good arguments, the foundation on which we lay our buildings. Land denotes place or context—it means the nation or the region we belong to, as well as the part of it that belongs to us; it is also a place of safety—we long for dry land, look for a landing place.

The Sacred Soil

Dakota children understand that we are of the soil and the soil of us, that we love the birds and beasts that grew with us on this soil. A bond exists between all things because they all drink the same water and breathe the same air.

—LUTHER STANDING BEAR, *My People the Sioux*

Conservation is a state of harmony between men and land. By land is meant all the things on, over, or in the Earth. Harmony with land is like harmony with a friend; you cannot cherish his right hand and chop off his left. That is to say, you cannot love game and hate predators; you cannot conserve the waters and waste the ranges; you cannot build the forest and mine the farm. The land is one organism.

—ALDO LEOPOLD, *A Sand County Almanac*

Life From Clay

The very ground beneath our feet may have been the catalyst that sparked the synthesis of life. The clay theory of the origin of life proposes that simple organic molecules arose by first binding to non-organic material such as clay particles. Clay essentially served as the matrix or skeleton around which simple organic molecules formed, attracted and bound to the clay by a tiny charge.

As an example, clay will concentrate amino acids out of solution. These molecules, which are the "beads" that, when strung together, form a protein "necklace," concentrate on the clay surface. Metals on the clay serve as catalysts to link the simple molecules into chains—more complex organic molecules called polymers. The clay provides a template or lattice, a launching pad of sorts, for organic molecules that might eventually migrate off the clay.

In one study, researchers found that the structure of clay was conducive to harbouring and protecting simple organic molecules. In this model, the molecules didn't cling to the outside of the clay; rather, they slipped between the sheets of minerals that made up the clay. The clay offered a protected haven for the organic molecules, which the researchers described as a sort of "primordial womb." Over time, the organic molecules could be expelled from the clay after heat and pressure changed the structure of the clay, closing up the space between the sheets. □

For most Indigenous people, land has been the foundation of life and the source of inspiration, identity, history and meaning. Kayapo leader Paiakan once described the land as "our supermarket and our drugstore." For most of human existence, we were nomadic hunter-gatherers who constantly moved as we foraged, so the concept of land ownership was an alien notion. Those ancient wanderers believed that they had the right to use the land as well as a responsibility to take care of it. This attitude persists to the present among Indigenous Peoples. Delgam Uukw, a contemporary Gitxsan and hereditary chief, said in 1987 in court on behalf of a major land claim:

> Each Chief has an ancestor who encountered and acknowledged the life of the land. From such encounters come power. The land, the animals, and the people have the spirit—they all must be shown respect. That is the basis of our law.

Thus, the human relationship with land requires that people protect it and maintain its fecundity. Their role was not to take too much but to leave some for others or another time and to return the remnants of their hunting or gathering back to the Earth. According to Hopi leaders:

> Hopi land is held in trust in a spiritual way for the Great Spirit, Massau'u... This land is like the sacred inner chamber of a church—our Jerusalem...
>
> This land was granted to the Hopi by a power greater than man can explain. Title is invested in the whole make-up of Hopi life. Everything is dependent on it. The land is sacred and if the land is abused, the sacredness of Hopi life will disappear and all other life as well...
>
> We received these lands from the Great Spirit and we must hold them for him, as a steward, a caretaker, until he returns.

This Indigenous sense of responsibility for the land is echoed in a document signed by such eminent scientists as Carl Sagan, Stephen Schneider, Freeman Dyson, Peter Raven and Stephen Jay Gould. It is remarkable for its use of a word such as *creation* in a spiritual sense and in its clear condemnation of the ecologically destructive path we are on:

> The Earth is the birthplace of our species and, so far as we know, our only home... We are close to committing—many would argue we are already committing—what in our language is sometimes called Crimes against Creation.

The Hidden World of the Soil

Soil offers far less to attract our attention than a marsh or tidal pool. Close inspection might reveal twigs, pebbles, perhaps a worm or a beetle and a matrix of tiny particles of sand. But the microscope exposes a far richer world, a place of ancient alchemy where hard and soft, liquid and gaseous combine, and where organic and inorganic, animal, vegetable and mineral all interact. Petals, leaves and stems fall from a plant to become compost for the seeds of that plant—death turns into life, grows up, feeds life and dies again, returning to the workshop underground to be restored to life. Almost all of the nitrogen that is essential for life must be made available through the action of nitrogen-fixing micro-organisms, most of which are in the soil. The soil is a microcosm where all the relationships of the larger world play out; in this element, earth, the other three unite—air, water and energy together create the vitality of the soil. Every cubic centimetre of soil and sediment teems with billions of micro-organisms; the soil produces life because it is itself alive.

Ironically, the money and vision expended on probing the secrets of Mars—$820 million for the latest two rovers alone—vastly exceed *what has been spent exploring the earth beneath our feet. Yet it is the soils of our gardens, fields, pastures, and forests, as well as the sediments beneath streams, lakes, marshes, and seas, that harbor the most diverse and abundant web of life known in the universe.*

—YVONNE BASKIN, *Under Ground: How Creatures of Mud and Dirt Shape Our World*

Soil organisms comprise a major portion of the total diversity of life. In fact, it is estimated that two-thirds of Earth's biological diversity lives in soil and underwater sediments. In this dark, teeming world, minute predators stalk their prey, tiny herbivores graze on algae, thousands of aquatic micro-organisms throng a single drop of soil water, and fungi, bacteria and viruses play out their part on this invisible stage. In their life and death these organisms create and maintain the texture and fertility of the soil; they are caretakers of the mysterious life-creating material on which they, and we, depend absolutely.

We know more about the movement of celestial bodies than about the soil underfoot.

—LEONARDO DA VINCI

What was true in Leonardo's time remains true today, in spite of all the advances in soil science over the past four hundred years. We know that a staggering number of organisms inhabit the soil (Table 4.1), but most of the species that have been identified remain virtually unstudied. A mere four thousand species of bacteria are formally recognized, for instance, yet, as soil ecologist Elaine Ingham explains, one *teaspoon* of forest soil can contain a billion individual bacteria, split among forty thousand species. The bacteria share that teaspoon with twenty thousand species of fungi, which, when stretched out, could extend for 150 kilometres. In healthy agricultural soil, there are about

|||||||||||||||||||||||||||||||||||||||||||||||||||||||||||||||||||||||||||||||||||||||||||||||||||

Table 4.1: Relative Number of Soil Flora and Fauna in Surface Soil

ORGANISMS	NUMBER/METRE^2	NUMBER/GRAM
Microflora		
Bacteria	10^{13}–10^{14}	10^8–10^9
Actinomycetes	10^{12}–10^{13}	10^7–10^8
Fungi	10^{10}–10^{11}	10^5–10^6
Algae	10^9–10^{10}	10^4–10^5
Microfauna		
Protozoa	10^9–10^{10}	10^4–10^5
Nematoda	10^6–10^7	10–10^2
Other fauna	10^3–10^5	
Earthworms	30–300	

600 million bacteria in a teaspoon, along with ten thousand protozoa (single-celled organisms such as amoebae) and twenty to thirty beneficial nematodes. The majority of these organisms have not been identified by scientists. A generous estimate is that perhaps 5 per cent of the species of soil animals have been identified.

Despite their microscopic size, soil micro-organisms are so abundant that they make up a significant biomass; in fact, they may be the major life form in any given area (Table 4.2). What's more, life even exists *below* the soil, in the very bedrock of our planet. Scientists had long assumed that life is confined to the four layers of soil above bedrock. But persistent reports that bacteria were contaminating drill bits in oil wells hundreds of metres deep prompted scientists to investigate. To their surprise, micro-organisms do exist, not just hundreds of metres, but several kilometres below Earth's surface, where temperatures exceed 50°C. Biologist Tullis Onstott found populations of bacteria embedded in the walls deep in the diamond mines of South Africa, for example. Here, in an environment of searing temperatures and crushing pressure, new microbes unlike any growing on Earth's surface were uncovered. Devoid of sunlight and oxygen, the energy sources for most life on Earth that we know about, these microbes might even be nuclear powered, fuelled by a reaction between water and the nuclear radiation emitted from the rocks.

The Living Underworld

If life underground is distributed in the density it is found in mine shafts, then the weight of protoplasm underground will be greater than the weight of *all* living things (whales, forests, herds of mammals and more) above ground. We know virtually nothing about this living underworld—how many species there are and how they are distributed, how they interact with other life forms, how they are involved with the movement of heat and nutrients from the planet's core, or even how they contribute to soil. Yet today, in order to reduce CO_2 emissions contributing to climate change, "carbon sequestration" is held out as a solution. The proposal is to pump hundreds of millions of tonnes of carbon dioxide into the ground where we have no idea how it is held, what kind of bonding occurs, where it is retained, how long it will remain underground and what it will do to the life forms that exist there. As creatures that live on Earth's surface, we regard the dark underworld as a lifeless, uniform matrix that we can use to sequester carbon without concern. It is time we learned to recognize our ignorance and acquire the humility to be very careful when we don't know enough. □

Table 4.2: Biomass of Soil Animals Under Forest and Grassland Cover

GROUPS OF ORGANISMS	GRASSLAND	OAK	SPRUCE
Herbivores	17.4	11.2	11.3
Dentritivores—large	137.5	66.0	1.0
Dentritivores—small	25.0	1.8	1.6
Predators	9.6	0.9	1.2
Total	*189.5*	*79.9*	*15.1*

Onstott said that embedded in the Earth's crust are thriving communities of bacteria with DNA signatures, that don't closely correlate to anything that's ever been found on Earth's surface. "So here we're standing essentially with new branches to the tree of life, some of them quite distinct from anything that's ever been discovered before." So while continental plates moved and crashed together, mountains rose and wore down, oceans filled and emptied, warm and cold periods came and went, these organisms of the deep underground carried on, perhaps dividing every millennium or so. These discoveries, and those of other extremophiles surviving in physically extreme environments such as hot springs or deepwater hydrothermal vents, demonstrate life's ability to grab hold and thrive, even in Earth's most challenging of places. Indeed, on Earth, even rocks can be alive.

The inhabitants of the soil play many different parts in its cycle of fertility. Larger organisms tunnelling through the soil help to introduce water and air and to mix minerals and organic materials through the layers of the soil, adding their droppings and eventually their bodies to the nutrient mix. Smaller organisms act as composters, breaking down and recycling organic matter and releasing nutrients for renewed growth. They fix crucial elements in forms that plants can use, and they interact with plants in many of the processes of growth. Worms, ants and termites, springtails, protozoa, fungi, bacteria—from the visible to the unimaginably minute, they are part of the crucial functions performed by the soil of the planet. More than just the substrate for creating growth, the soil is Earth's primary filter, cleansing and recycling water and decaying material; it is also a major component of the planet's water-storage and water-cycling processes.

Perhaps our most precious and vital source, both physical and spiritual, is the most common matter underfoot which we scarcely even notice and sometimes call "dirt," but which is in fact the mother-lode of terrestrial life and the purifying medium wherein wastes are decomposed and recycled, and productivity is regenerated.
  —DANIEL HILLEL, *Out of the Earth*

The Origins of Soil

The planet's terrestrial region wasn't always the familiar soil-encrusted surface that nourishes us. Soil is a complex mixture of mineral particles, organic material, gases and nutrients. How these ingredients combine to make soil is part of the long story of the evolution of Earth, a story of making and remaking in which, once again, life itself plays a starring role.

Soil begins when rocks are broken down by weathering—the powerful assault of natural forces. Thus, the kind of soil in an area depends to some extent on the minerals contained in local rocks. The most abundant mineral on Earth is feldspar, a group of crystalline minerals that are an essential component of nearly all crystalline rocks. In Earth's earliest days, as the planet cooled and solidified, the feldspars, which melt at temperatures as low as 700° to 1000°C, soon melted and rose to the surface, where they are extremely common, and eventually became the mineral source of clay.

Rock formed by cooling magma—molten rock below the surface of the Earth—is called igneous rock (the other two types of rock are sedimentary and metamorphic); at the surface, it is exposed to constant weathering. Changing temperature, wind, rain, snow and humidity break down rock; the resultant material is moved across the surface of the Earth by gravity, running water, glaciers, wind and waves. Only two hundred years ago, it was thought that mountains, lakes and deserts were permanent and unchanging. Now we know that the surface of the globe has been undergoing constant change since its beginning—and that it still is; even mountains can wear down over time through weathering and erosion. (Mountains can grow, too. At present, uplift in the Himalayas is about 1 centimetre each year.)

To grasp the forces of weathering, think of a newly laid concrete sidewalk. As time and people pass, that smooth, impeccable surface will be roughened; wear and tear will pit the surface and stones will begin to appear; tree roots will heave and crack the concrete; water draining from driveways beyond will erode channels. Imagine people's feet wearing away the surface particle by particle, dust-laden winds scouring the surface, winter rain infiltrating a thousand tiny pores and expanding as it freezes. In many ways rock is similarly affected over enormous periods of time.

There are three types of weathering: mechanical, chemical and biological. Mechanical weathering breaks rock into smaller fragments without altering its chemical structure. Water percolates into cracks and crevices; expanding as it freezes, it can split the rock apart. Rocks exposed by erosion of material around them break and slide down the hill, smashing other rocks in the process. As temperatures rise and fall between day and night, rocks expand and contract, forming fissures under the repeated strain. The slow, repetitive dripping of water over the ages can wear away a mountainside.

Chemical weathering alters the chemical structure of rock by removing some constituents and leaving others; this process too is part of soil formation. Water is often the agent of chemical weathering. For example, water can dissolve pyrite and sulphite minerals to form highly reactive sulphuric acid. Some of the dissolved products become potential nourishment for life or react with minerals in rock. Water naturally picks up carbon dioxide as it falls through the sky and percolates through the soil. This process results in water that is slightly acidic; in fact, it forms weak carbonic acid. Given enough time, water has the power to dissolve almost anything. Water can break down feldspar to form clay and grains of sand. The steady work of water etching away limestone has resulted in the world's most outstanding cave systems such as New Mexico's Carlsbad Caverns or the elaborately riddled network of chutes, tunnels and sinkholes in the limestone of Jasper National Park's Maligne Canyon.

When rock breaks up, more surface area is exposed to potential chemical weathering, which decomposes the constituents of this rock into new compounds (Figure 4.1). Mechanical and chemical weathering can produce a gram of clay with particles having a total surface area approaching 800 square metres, a large reactive area analogous to the surface of the alveoli of lungs and intestinal villi.

During the early days of weathering on this planet, calcium silicates in basalt were dissolved completely in water, producing calcium carbonates and silicic acid in solution, which eventually flowed out to the oceans and ended up on the bottom as sediment. A mix of elements began to build in the sterile oceans; this mix would transform the Earth.

Once life had evolved, living organisms introduced a third type of weathering: biological weathering. Micro-organisms

||||||||||||||||||||||||||||||||||||||||||||||||||||||||||||||||||||||||||||||||||||||||||||||||||||||||||||

Figure 4.1: The increase in surface area after fragmentation.

Adapted from Frank Press and Raymond Siever, *Earth* (San Francisco: W.H. Freeman and Company, 1982), p. 92.

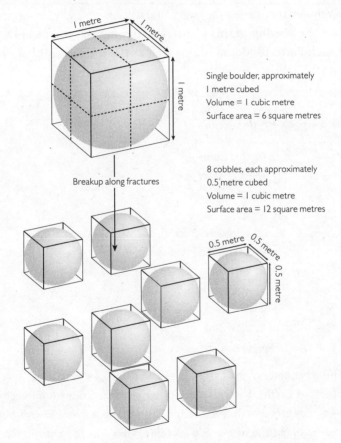

Single boulder, approximately
1 metre cubed
Volume = 1 cubic metre
Surface area = 6 square metres

Breakup along fractures

8 cobbles, each approximately
0.5 metre cubed
Volume = 1 cubic metre
Surface area = 12 square metres

extruded chemicals that leached rock for useful elements. Infiltrating into cracks, masses of organisms exerted pressure, as did the roots of plants seeking minerals and a hold on rock faces.

Weathering Into Life

The intensity of weathering depends on climate, tectonics, the original composition of the rock, and time. Mechanical and chemical weathering produced clays, soils, sediments and salts in the oceans, while breaking apart of rock created boulders, pebbles, sand and silt. These processes may seem at first to be massively destructive, but they were, and still are, crucially important in the flowering of this planet. Weathering was the means by which Earth produced life, as well as the instrument that life

The Greek Myth of Gaia

> *Mother Earth*
> *The mother of us all,*
> *the oldest of all,*
> *hard,*
> *   splendid as rock*
> *Whatever there is that is of the land*
> *   it is she*
> *   who nourishes it,*
> *It is the Earth*
> *   that I sing.*
> —HOMER

The great deity of the early Greeks was Gaea, or Gaia, the "deep-breasted Earth." She was the mother goddess from whom everything else came forth. She created Uranus, the starry sky; together they peopled the new-made universe. Gaia was the cosmic creator: from her came the Titans, the first race of gods, the giants, the storm spirits and all the other forces of nature, as well as human beings. "Men and gods," wrote the poet Pindar, "we are of the same family, we owe the breath of life to the same mother."

used to create a new medium on which it could thrive and diversify.

This is how scientists tell that extraordinary story. Dissolved elements and salts in the oceans formed a rich broth of atoms and molecules; they were the precursors of life. Long before the land surfaces of the planet became hospitable, conditions in the oceans rapidly evolved to make life inevitable. The first cells that arose more than 3.6 billion years ago were bacteria; for half of the time that has elapsed since then, the only forms of life that

In later myths one of Gaia's descendants, Demeter, took on part of the great Earth mother's role as the guardian of cultivated soil, fertility and the harvest. The story of Demeter's loss of and reunion with her daughter, Persephone, dramatizes the natural cycles of the Earth as the source of life, renewal and regeneration in the soil. As Persephone gathered flowers in a sunlit field one day, she was kidnapped by her uncle Hades, king of the underworld. Hades's dark kingdom was the fearful destination of the dead, but it was also the source of life and growth. When Persephone disappeared underground, all growth stopped on Earth as Demeter searched the world for her lost daughter. Eventually the gods returned her to her mother so that harvests could resume, but not before the girl had eaten some pomegranate seeds in the kingdom of the dead. Those seeds condemned her to return beneath the ground for part of every year.

All growth, every harvest, comes from the dark, rich realm of the soil; new life is nurtured by death, by the decaying leaves and plants, the accumulated remains of innumerable generations of organisms. Every year the world must die to be reborn again from the earth—from soil and from planet—as Homer writes, from the mother of us all. □

existed anywhere on Earth were bacteria. Further evolution required a substrate: the soil. Just as living organisms had interacted with and profoundly influenced the makeup of the atmosphere and water, so living organisms altered the planet forever by creating soil. Bacteria moved onto the land and began to dissolve the barren rock—taking it apart to obtain the nutrients they needed. They were the vanguard invasion of life on dry land.

When plants finally moved out of the water 350 million years ago, they had to gain a foothold on solid rock or seek out bacterium-encrusted gravel or sand. Plant roots reached out to find purchase in any crevice or hole, secreting enzymes that dissolved rock to provide essential elements. But it was the bacteria that excelled at exploiting the hard land surface, dissolving and concentrating substances they found ways to exploit. They evolved ways to create carbonates, phosphates, silicates, oxides and sulphides. In this way, both bacteria and plants broke boulders apart. Over immense periods of time bacteria and plants, combined with other forms of weathering, reduced mountains to rubble. Plant roots held together grains of sand and protected them from the wind and the rain. As countless generations of animals, bacteria and plants died, their carcasses were added to the sand. As their tissues and cells broke down, molecules were liberated to form the organic matter that builds that rich brew of crumbly, gritty, brownish *stuff* that provides habitat and nourishment for life on Earth.

Over millions of years, micro-organisms and plants crept farther and farther onto land. Drawing sustenance from the sun, water and minerals, they spread across shores, along valleys and onto plains. As centuries became millennia and then hundreds of millennia, the organic remains of plants piled up, providing a matrix on which bacteria flourished. Eventually, small animals—worms, arthropods—found ways to exploit the growing soil, accelerating the breakdown of rocks and minerals with their burrowing and adding their carcasses to the composting materials.

As soil became richer and deeper, plants became larger and more abundant. Decaying organisms helped to hold rock and particles in soil material that was rich in organic matter. As continents were covered with this stockpile of complex molecules found only in living organisms, plants flourished,

The Balance Between Life and Death

During the 1970s, after the Arab oil embargo had stimulated an urgent search for domestic oil, I was working on a film about the hazards of drilling for oil in a frontier area above the Arctic Circle. Walking on an Arctic island on rock-strewn soil stained by lichen, I spotted an incongruous patch of blazing colour. It was a miniature garden of small flowers and grasses with a cloud of insects hovering around it. In the centre of this tiny Eden were the crumbling bones of a muskox that had died many years before. Bleached by the sun and brittle from the leaching of its minerals, the bones still had enough nutrients to support this community of creatures in the unforgiving environment of the Arctic.

At the other end of the world, I visited an old whaling station in Antarctica. Amid the rusting vats used to render the blubber and the bleached, rundown buildings that had housed the whalers, plants and flowers grew on the blood and bones of countless leviathans beached on the barren sand. And in northern Europe, the blood and bone left in the aftermath of the terrible battles of the First World War continue to nourish the vast poppy fields that are literally the reincarnation of generations past.

All of these examples demonstrate the exquisite balance between life and death, one cycle emerging from the other. □

providing habitat and food for a growing menagerie of terrestrial animals, of which human beings were the most recent
addition.

Today soil is a complex and diverse mix, varying wonderfully
from place to place and supporting a multitude of communities.
In the rainforests of the planet, high above the surface of the soil,
immense trees strain toward the sun, forming a dense canopy
filled with other forms of life. Those plant colossi are held aloft by
anchors of serpentine roots weaving along the tropical soil that

Networking for Life

Fungi are often portrayed as agents of decay and decadence in natural systems. Fungi are critical to the recycling of organic matter,
but they are also unlikely allies in the overall health and robustness
of ecosystems. Towering forests and prairie grasslands—in fact,
almost 90 per cent of all plant species—are supported by their
association with soil fungi.

In Europe, truffles have long been delicacies to the highest epi-
curean level. Since truffles exude a compound resembling male
swine pheromones, female pigs are used to detect the odorifer-
ous delicacies in the forlorn hope of a sexual encounter. In the late
nineteenth century, the king of Prussia decided to wrest some of
the truffle trade away from France by finding a way to culture the
mushrooms. He assigned the task to mycologist A.B. Hatch, who
painstakingly traced the hyphal strands extending from the fruiting
body, only to find that they invariably wrapped themselves around
the root hairs of trees in such intimacy that they seemed to be a
natural part of the tree. ("Mushrooms" that poke through the soil
are the fruiting body of a fungus—they carry the spores. Most of
a fungus's biomass comes from a mesh of tiny rootlike hyphae that
thread through the soil.) He had discovered a remarkable associa-

is infertile, owing to intense weathering and leaching, or plunging deep into the earth in temperate regions.

On the diverse soils of prairie grasslands, Arctic wetlands or equatorial savannahs, mosses, grasses, shrubs and flowering plants support each other, as well as many different species of flying and creeping insects and multitudes of herbivorous mammals. The great caribou herds of northern Canada, the wildebeest of the Serengeti, the vanished millions of prairie bison—these are just some of the soil's teeming family.

tion of enormous significance, the mutually beneficial association of mycorrhizal fungi and trees. Around their roots, each tree may have dozens of different mycorrhizal fungal species, some specific to that tree species, others not.

As many people know from the writing of Suzanne Simard and Peter Wohlleben, plants and fungi support one another in a mutually beneficial, symbiotic relationship. For the plant, mycorrhizae increase the absorptive surface of its roots, facilitating a greater uptake of water, minerals and nutrients such as nitrogen and especially phosphorus. In essence, the threadlike mycelia act as extensions of the plant's roots, infiltrating a far greater area of soil than the plant could manage on its own. As part of the deal, the fungus also stimulates growth and produces antibiotics that can protect the plant from pathogens. In return, the fungus benefits by receiving a steady supply of sugar its host synthesizes in its leaves.

So critical is the role of this co-operative relationship that it is widely recognized that plants grow poorly, or not at all, in soils without the mycorrhizal fungi. In fact, it is thought that it is because of this relationship that plants were able to colonize land so successfully. The fossils from some of Earth's earliest plants show evidence of mycorrhizae. □

The Deep Horizons

Soil has been called the bridge between life and the inanimate world. The bridge is built, as we have seen, out of minerals and organic material, but it also requires air and water. Soil acts as a reservoir for carbon and thus to a large extent controls the partitioning of terrestrial carbon and atmospheric carbon. Half of the volume of good-quality surface soil is a mix of disintegrated and decomposed rock and humus (decayed remains of animal and plant matter); humus enhances the ability of the soil to hold water (Figure 4.2). The remaining half is composed of pores where air and water circulate. The pores are as vital as the solid areas because circulating air and water provide oxygen and carbon dioxide for micro-organisms and plants.

The great trees of the forest can be seen as one of the ways in which the inanimate and the animate worlds meet in the soil. The

IIIIIIIIIIIIIIIIIIIIIIIIIIIIIIIIIIIIIIIIIIIIIIIIIIIIIIIIIIIIIIIIIIIIIIIIIIIIIIIIIIIIIIIIIIIIIIIIIIIIIIIIIIIIIIIIIIIIIIIIII

Figure 4.2: Profile of fertile soil.

Adapted from Frank Press and Raymond Siever, *Earth* (San Francisco: W.H. Freeman and Company, 1982), p. 124.

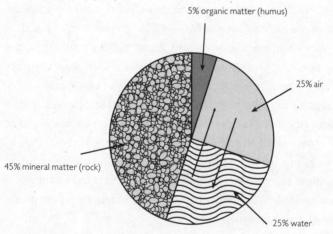

5% organic matter (humus)

25% air

45% mineral matter (rock)

25% water

roots of these trees extend laterally in an ever-branching network of successively finer and finer filaments that probe the soil, foraging for moisture and nutrition. In the soil, moisture is held in the matrix of soil particles and moves very slowly, so a plant's roots are constantly on the move trying to find it. The total length of

||||||||||||||||||||||||||||||||||||||||||||||||||||||||||||||||||||||||||||||||||||||||||||||||||

Figure 4.3: The four horizons of soil.

Adapted from Frank Press and Raymond Siever, *Earth* (San Francisco: W.H. Freeman and Company, 1982).

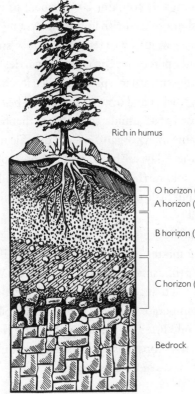

Rich in humus

O horizon (organic material)
A horizon (minerals and humus)

B horizon (fine clays)

C horizon (weathered bedrock)

Bedrock

the root system of an individual tree can be several hundred kilometres, with a surface area of several hundred square metres!

Soil is formed from the surface downward, and that results in variations in composition, texture, structure and colour. Because differences can be observed at varying depths, soil can be defined as a series of zones, layers or horizons. While soil scientists identify the layers in different ways, four are commonly defined from top to bottom as O, A, B and C (Figure 4.3).

The O horizon is largely organic material. The upper part contains primarily recognizable plant litter, such as leaves and stems, whereas the lower portion is made up of partially decomposed organic matter and humus. The A horizon is largely composed of minerals and humus. It contains a great deal of biological activity and is much more fertile than the underlying soil. This layer becomes coarser in its lower regions because small particles wash out as water percolates through and carries dissolved inorganic materials downward in a process called leaching. The B horizon is generally thicker and overlies weathered bedrock. The B horizon is composed of accumulated fine clays that pack down as hardpan; it contains living organisms and organic matter but much less than in the A layer. The O, A and B horizons constitute true soil, or the solum. The C horizon is made up of partially altered rock debris but little organic matter, which is the crucial component for fertility.

In nature, the difference between the diners and the dined upon may be just a matter of time. Life thrives on life; it is a rule of nature. Over thousands of years, this process generated organic material in soil that accumulated relentlessly but almost imperceptibly. In temperate climates where deciduous forests and annual plants pile litter on the forest floor each year, the average rate of growth in topsoil is 5 centimetres per millennium. This is the stuff that makes our planet fruitful.

A land of wheat and barley, vines,
    fig trees and pomegranates,
A land of olives and honey;
A land wherein thou shalt eat
    bread without scarceness.

—DEUTERONOMY 8:8–9

Feeding From the Soil

Air is the breath of life, water the drink and the earth the food of life. We do not, of course, eat earth directly, but we absorb it every day of our lives as a matter of necessity. The green Earth is the meadow we graze in, the ground we are shaped from, the daily bread that keeps body and soul together.

Imagine Earth as an orb with a diameter of 70 metres and covered in a skin the thickness of that of a tomato. That thin outer layer corresponds to the fine wrapping of soil that covers the surface of our immense planet. The constant renewal of life on Earth occurs in that thin layer; we, like all other terrestrial life forms, depend on it, directly or indirectly, for our food.

Our daily survival and well-being depend on the consumption of sufficient amounts of energy, proteins, carbohydrates, inorganic substances, trace elements, essential fatty acids, vitamins, water and roughage (non-digestible constituents of plants such as cellulose). Every day fat provides 25 per cent of our energy needs, protein 12 per cent and carbohydrates 63 per cent. Today much of our food is processed and put into packages that disguise its biological origins, but every bit of what will be incorporated into our bodies, except for chemically synthesized molecules such as sweeteners and fat alternatives, comes from pre-existing life; much of it comes directly from the soil.

Vitamins are a special reminder of our dependence on other life forms. Vitamins are complex biological compounds that are necessary for human physiology but that can't be synthesized within our own bodies. They are made by other species, which we eat and from which we obtain the essential vitamins. Without vitamins, we suffer documented problems including night blind- ness (caused by the lack of vitamin A), scurvy (lack of vitamin C), rickets (lack of vitamin D), megaloblastic anemia (lack of folic acid), pernicious anemia (lack of vitamin B), beriberi (lack of vita- min B), pellagra (lack of nicotinic acid) and blood-clotting distur- bances (lack of vitamin K).

Our bodies are wonderfully adapted to obtain the nutrition we need. Like the creation of soil, feeding is a matter of taking things apart in order to put other things together. You might think of digestion as another version of weathering; we perform heroic acts of physical and chemical weathering on our foodstuffs to break them down and use them in our own vital processes.

Most of the process of absorbing food is done by reflex, with- out conscious thought; like breathing, eating is so central to our existence that our bodies are designed to bypass the conscious- ness and get on with the job.

Digestion begins right in the mouth when we insert food past our lips and start to tear and chew it into small pieces (Figure 4.4). Consisting of hard enamel over bone-like dentine that cov- ers nerve- and blood-rich pulp, the thirty-two teeth of a human adult are engineering marvels. Chisel-like incisors cut the food, while flat-topped molars grind it down.

Receptors reacting to smell, taste and touch, as well as to the act of chewing itself, stimulate the production of saliva from var- ious glands. Saliva is a great deal more than just spit. It contains mucus, which moistens the food so that it can be more easily moved through the mouth for mastication and swallowing, and the enzyme alpha-amylase, which breaks down carbohydrates. Saliva also contains immunoglobulin A, lysozyme and peroxidase, powerful agents that combat infection from viruses, bacteria and other pathogenic organisms.

Swallowing is no simple matter, but we achieve it in style by means of a series of reflexes. Chewed food is collected into a lump called a bolus and propelled by the tongue into the pharynx at the top of the esophagus. When the bolus touches the pharynx,

Figure 4.4: The components of the digestive system.

Adapted from Cecie Starr and Ralph Taggart, *Biology: The Unity and Diversity of Life*, 6th ed. (Belmont, CA: Wadsworth, 1992), fig. 37.4.

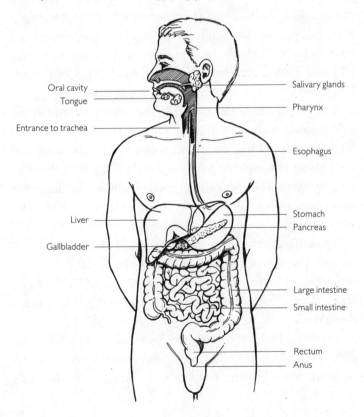

sensory receptors stimulate a sequential wave of contractions called peristalsis. This squeezes the bolus and pushes it down to the sphincter that controls entry into the stomach.

Food passing through the sphincter distends the stomach, causing it to contract and push the bolus toward the "bottom" half of the stomach. Hormones regulate the rate of contraction. In ten to twenty minutes, half of the water in the food has been absorbed by the stomach lining. Stimulated by sight, aroma and taste, the stomach daily secretes about 3 litres of gastric juices: pepsinogens for breaking down protein, mucus to protect the stomach lining from digestive juices, and hydrochloric acid and gastroferrin to absorb iron. The food is converted by chemical and mechanical action into a liquefied state called chyme, which is pushed toward the bottom of the stomach, compressed and then propelled back by the contractions of the stomach. This back-and-forth action serves to increase liquefaction, mix gastric juices, partially digest the chyme and emulsify fats.

The chyme then passes out of the stomach through the pyloric sphincter into the small intestine, which is about 2 metres long. Here, the absorption of breakdown products, water and electrolytes is completed. Like the alveoli of the lung, the intestines have a series of intricate outpockets, which form finger-like projections called villi and microvilli to maximize surface contact with the chyme (Figure 4.5).

The surface area of the villi and microvilli exceeds 100 square metres. The tips of the villi constantly slough off as they are moved back and forth by muscular contraction. Waves of peristaltic contractions in the intestine push the mixture of breakdown products, water and electrolytes toward the anus.

Every day the pancreas releases about 2 litres of juice rich in carbonic acid to neutralize the chyme, and digestive enzymes to break down protein, fats and carbohydrates. To digest fat,

III

Figure 4.5: Fine structure of villi and microvilli.

Adapted from Cecie Starr and Ralph Taggart, *Biology: The Unity and Diversity of Life*, 6th ed. (Belmont, CA: Wadsworth, 1992), fig. 37.9.

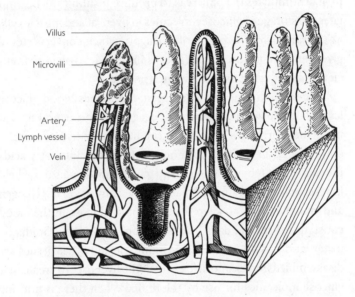

the liver produces about 0.7 litres of bile every day; some of it is stored in the gallbladder. About 85 per cent of the bile is bilirubin, which is the degradation product of hemoglobin from dead red blood cells. The bilirubin breakdown products that are excreted through the intestines are what make feces brown.

The last part of the gastrointestinal tract is about 1.3 metres of colon, or large intestine, where water and electrolytes continue to be removed. At the end of this tortuous journey, what remains · is feces—mostly unabsorbed vegetable matter, sloughed-off cells, bacteria and some water. There are $10^{10}–10^{20}$ bacteria per millilitre in the adult colon, and they make up a third of the dry weight of feces.

If we have fed well, amino acids, simple sugars and fatty acids enter the bloodstream as digestion proceeds. Glucose, one of the sugars, acts as a regulatory mechanism. When blood glucose is high, it stimulates the pancreas to produce insulin, which stimulates the storage of glucose molecules in liver, fat and muscle cells. At the same time, secretion of glucagon, which converts stored glycogen into glucose, in the liver is inhibited. So glucose is efficiently stored for use when energy is needed.

Under conditions of fasting or starvation, as blood glucose levels fall, the adrenal gland produces epinephrine and norepinephrine, hormones targeted to the liver, fatty tissue, muscles and other tissues. These hormones convert fats into fatty acids for energy, thereby allowing many glucose molecules to be held in reserve for the brain. By balancing the conversion of glycogen and fats, the liver can keep the amount of glucose in the blood constant for several days before it drops drastically. Gradually, if starvation continues, proteins in muscle and other tissues are disassembled into amino acids, which the liver uses to create new glucose molecules for use by the brain. When there is nothing coming in from outside to be taken apart by digestion, the body takes itself apart, sacrificing tissue to keep the command organ, the brain, working as long as possible.

We are constantly wearing out and using up molecular components of our tissues and organs. Blood cells, for example, last only 120 days; epithelial cells lining the gut, five days; and the epidermal cells covering the skin's surface, about two weeks. Cells die by the millions each day and must be replaced. Our bodies are in a constant and steady state of death and renewal—even our skeleton is replaced in its entirety every few years. That is why our need for nutrition is absolute. Since all of our food was once living, those plants and animals absorbed whatever was in the air, water and soil. Throughout the history of life on Earth the excrement or waste of one species has been a resource for other species in an endless cycle of use, elimination and reuse. Like air and water, soil is a critical source of life that is taken into the deepest recesses of our bodies and actually made into us. Like air and water, then, soil demands great respect: what we do to it, we do to ourselves.

Agriculture: A New Phase in Human Development

Some ten thousand to twelve thousand years ago, human beings realized that seeds placed on or in soil would grow into plants that were useful to people. This realization led to the Agricultural Revolution, which fundamentally altered human behaviour and provided the foundation of civilization. Early farmers generally worked within the constraints of their environment. To feed themselves, they needed to understand the vagaries of the land, the climate and the plants they grew. They learned to recognize the differences between soils, and, by trial, error, observation and experience, generations of early farmers domesticated plants with the qualities—such as size, taste and sugar, fibre or oil content—they desired. By steering the selection of plants that fit their needs—and those of their environment—these first farmers created resilient, locally adapted food crops that fed the world for centuries: rice and millet in China, sorghum in Africa, and corn and squash in the Americas, for example.

In domesticating plants and then animals, farmers freed humanity from the exigencies of nomadic hunting and gathering, providing a basis for permanent settlements in which individuals could develop specialized roles in an increasingly complex society. Since farming now produced more food than people needed, food could be stored for later use or traded and sold for other goods. In the burgeoning settlements, people began to develop skills that would both facilitate agriculture and the growth of their communities—carpentry, toolmaking, pottery and weaving, for example. People were no longer on the move, and they began to accumulate material goods.

As we became expert at feeding ourselves, that expertise allowed our numbers to grow and our societies to change. Hierarchies established a division of labour: there were those who owned land and those who worked it. Over time this social structure led to competition within communities and between societies over land and water.

With each new innovation and invention, agriculture and societies shifted. About seven thousand years ago, for example, the invention of a metal plow that could be pulled by domestic animals meant that plots could be larger and land that was previously too difficult to break by hand could now be cultivated. In arid regions, primitive irrigation systems—primarily diverted watercourses—were established. Agrarian societies slowly began to change the land that fed them.

The effect of human activity on soil did not begin solely with agriculture, however. Even in the preagricultural Paleolithic period, human beings armed with little more than stone spears and axes were able to extinguish large mammals in what is now North and South America. But it was the invention of agriculture, particularly modern industrial farming, that altered the rate and scope of change. Modern farming techniques, such as the use of heavy machinery, irrigation and the extensive use of chemical fertilizers, have had a devastating impact on the soil. Modern techniques have increased productivity per hectare, but the organic material produced is not returned to the soil to complete nature's cycle; instead, it often ends up in sewers, landfills or incinerators.

... civilized man has consistently tried to make the land fit his pattern of farming, when he should make the farming fit the pattern the land.
—VERNON GILL CARTER and TOM DALE, *Topsoil and Civilization*

In 1984, Canadian senator Herbert Sparrow's report *Soil at Risk* documented the rapid depletion of topsoil accumulated over millennia that has occurred in Canada within the past few decades, a pattern duplicated in the United States and Australia. The report suggested that the kind of factory farming that has become the norm in industrialized countries since the last war, with its excessive emphasis on productivity, has been chiefly to blame. The problems depicted in Sparrow's report have only become worse. Urbanization, along with poor practices in agriculture and forestry, has caused compaction, soil loss through erosion and decreases in organic matter, which in turn means less water enters the soil to be cleansed of pollutants. In addition, fast snowmelt and intense rainfall lead more frequently to flooding because thinner soil cannot retain as much water.

Civilized man has marched across the face of the Earth and left a desert in his footprints.
—ANONYMOUS

The incredible increase in Earth's human population in the past fifty years has been matched by a similar increase in agricultural productivity. But this productivity has been achieved by deliberately overriding the natural limits of the soil and the living system that makes the soil so productive. According to population biologist David Pimentel, modern farming methods deplete topsoil about thirty times faster than it can be replaced:

More than 99% of food worldwide comes from the soil ecosystem. Rapid erosion of soil is reducing food production—and causing serious losses in biodiversity. Over the past 40 years, approximately 30% of the world's cropland has become unproductive... which has contributed to the malnourishment of more than 3 billion people.

It takes an average of five hundred years to build 2.5 centimetres of topsoil. As Pimentel explained it, the global loss of topsoil now exceeds new soil production by 23 billion tonnes a year, which is 0.7 per cent of the world's soil. At that rate, more than 30 per cent of the world's current supply of topsoil will be gone by 2040, while human numbers and demand will continue to grow.

Earth's diverse living things cleanse, alter and regenerate air, water and soil—elements they help to create and depend on absolutely. Yet now one species out of untold millions has commandeered almost half of the planet's soil. According to Dr. Bernard Campbell, we humans "either use, co-opt or destroy 40 per cent of the estimated 100 billion tons of organic matter produced annually by the terrestrial ecosystem." By this means we drive many other organisms who are keeping the planet habitable to extinction.

Fire and Farming

After humans learned how to make fire, they deliberately burned forests, wetlands and grasslands and in the process caused soil erosion, slides and siltation. In Australia, when the ancestors of today's Aborigines arrived forty thousand to seventy thousand years ago, they transformed the plant and animal makeup of the continent with the widespread use of controlled burning. But fire should not necessarily be viewed as a destructive force.

In fact, fire is a natural, even necessary, phenomenon in many environments.

In dry climates, wildfires help reduce the "fuel load" in an ecosystem. Periodic fires are of relatively low intensity and tend to burn in patches, resulting in a mosaic of different-aged vegetation. Fires help break down the natural litter, releasing the nutrients back to the soil, thus stimulating plant growth. (Many plants are even fire dependent, requiring the heat of a fire to release seeds.) If fires are suppressed for too long, the intensity of the inevitable fire increases, resulting in fires that can burn to bedrock.

Traditional "firestick" farming of Australian Aborigines reduced the risk of catastrophic, uncontrolled fires. Regular and calculated use of fire was one way in which they managed their environment, opening up dense areas of bush for wildlife and reinvigorating plant growth for themselves and their prey.

The use of fire was central to the Aboriginal people. In the book *Nourishing Terrains,* author Deborah Bird Rose explains that the fires European explorers saw burning across Australia meant "everything was good—that people and country were doing the right thing."

> ... *fire burning country is not just fire, smoke and blackened vegetation. Fire country involves people who have ways of interpreting their place within the environment where they live, on the country they call home. Their relationship with fire at its most basic is as a tool, but fire is always related to events associated with the past and the future... Fire can be seen to be part of an ecology of internal relations; no event occurs which stands alone.*
> —JOHN BRADLEY, quoted in *Nourishing Terrains* □

We now know, alas, that nature is finite in her beneficence, and that
we have come close to her limits... Most parts of the world capable of
sustaining a high level of agricultural productivity are already devel-
oped—those areas not already developed (such as the Amazon basin) are
probably not suitable for it. So many apparently valuable development
projects... have had appalling and unexpected results.
—BERNARD CAMPBELL, *Human Ecology*

Soil: A Priceless Material

Soil continues to be the main source of humankind's nutrition.
Although throughout the world consumption of fish equals that
of all cows and chickens combined, most people in the world live
primarily on grain. Thus, agriculture provides more than 98 per
cent of human nutrition. Of the terrestrial portion of the planet,
12 per cent is cropland, 24 per cent is pasture and 31 per cent is
forests. National parks, which conserve global biological diversity, take up a mere 3.2 per cent of the total terrestrial ecosystem.
A third of the land on the planet is unsuitable for agriculture, forest or pasture.

The finite productive arable land is a slender base on which
to support humanity, and it is shrinking by more than 10 million hectares a year because of soil degradation. Adding to the
pressure from loss of productive land is the relentless increase
in population by more than 80 million a year, requiring another
10 million hectares, an area the size of the state of Ohio. That's
why deforestation is occurring at such a rapid rate; 80 per cent of
forest loss is to grow food. The most serious cause of soil loss and
land degradation is erosion.

An Old Wisdom of the Earth

Thou shalt inherit the Holy Earth as a faithful steward... Thou shalt safeguard thy fields from erosion.

—WALTER CLAY LOWDERMILK,
"Conquest of the Land Through 7,000 Years"

The interconnectedness of all things on Earth means that everything we do has consequences that reverberate through the systems of which we are a part. When we reclaim this ancient understanding, we will recover the sense of responsibility that it entails. The Waswanipi people, among other Indigenous groups, had a clear understanding of this responsibility:

> The traditional Waswanipi hunter says that success in hunting is not entirely his own doing. A successful kill can partly be attributed to the willingness of the particular moose or beaver or whitefish itself to lay down its life so that Waswanipi people can live...
>
> [They] know that the north wind and the souls of their prey are neither capricious nor passive but are a dynamic indication of the hunters' current moral standing in the "eyes of nature." The north wind and the animal spirits operate in a reciprocal relationship with the hunters' actions today and in the past.

In exchange for nature's generosity in meeting Waswanipi needs, the hunter acknowledges his profound obligation to act responsibly toward nature. His many responsibilities include killing animals swiftly, without causing unnecessary suffering; neither killing more than is given nor killing for fun or self-aggrandizement; acting respectfully toward the bodies and souls of the animals by observing the proper rituals when retrieving, butchering and eating them; and using what he is given completely, without carelessness or waste.

> Waswanipi hunters and people learn from their elders that the body of the animals the hunter receives nourishes him, but the soul returns to be reborn again, so that when men and animals are not in balance, the animals are killed but not diminished, and both men and animals survive.

Subsistence farmers show the same responsibility for the soil they till, feeding and tending their land so that it in turn will feed their families over the long term. In every region the lore of the land is a local compendium of wise ways to coax a harvest from the soil—when to plow, which plant species to combine, how to protect the soil from the weather so that the forces that put it

The Aral Sea Disaster

The village of Muynak was once a vibrant seaside resort and fishing community. Today, the Aral Sea is 160 kilometres away. A white dust, which, from a distance, looks like snow, coats the dry seabed. But it is not snow; stretched out from the village is a wasteland of salt, pesticides and heavy metal residues, which are regularly lifted into the wind in contaminated dust storms. These toxic winds blow across the exposed, scorched seabed and whistle through the rotting hulks of long-abandoned fishing vessels and on into the villages. Where there were once thriving, healthy communities, there remain villages with contaminated drinking water, unprecedented levels of asthma, tuberculosis and cancer, and an underemployed and despondent population.

The Aral Sea was once the world's fourth-largest lake, full of fish, and the lifeblood of nearby communities. The sea regulated the climate and supplied the river and ground water of the region. It supported over 173 species of terrestrial wildlife (even some tigers), and 24 species of fish swam in its waters. But in the 1950s, the Soviet Union envisioned the region as a place to grow cotton—a crop considered to be "white gold"—and diverted the flow of the rivers that fed the Aral Sea. In 1960, the Aral Sea was the size of Ireland. Now, after decades of water being diverted through an archaic and inefficient irrigation canal system, its eastern basin has completely dried up, forming the world's newest desert, the Aralkum Desert. More than 90 per cent of the sea has dried up since 1960, leaving only isolated bodies of water. Instead of rich soil, vast salt plains surround the sea, and the moderate climate has changed, with hotter summers and colder winters. Dust storms spread disease, and most of the land is so contaminated by salt and by pesticides used in the cotton industry that it cannot support life.

The Aral Sea is one of the most tragic environmental catastrophes of the last century. It has even been called "the quiet Chernobyl." But there is some hope for the region. After years of rehabilitation, including the construction of new dams, dikes and sluices to repair previous damage, the North Aral Sea is showing signs of improvement. One of the rivers, the Syr Darya, now flows freely, and the amount of water reaching the sea has doubled. Before the rehabilitation, the North Aral's main port city, Aralsk, was 80 kilometres from the sea. Now the shore is within 20 kilometres, and fish have started to return. Given time, the Earth can heal itself, but the disaster of the Aral Sea shows how quickly we can alter the nature of our environment and how one change can ripple outward with catastrophic and far-reaching effects. □

together cannot take it apart. The farmer must artfully fit human needs into the natural systems he (or, more usually, she, in most parts of the world) is tapping into—the local web of life, adapted to local conditions, designed by the place over a long period of time. It is only now in the "developed" world that we seem to believe we can improve on nature and rewrite the rules.

Technologically advanced nations have not been using the soil in a sustainable way; instead, they have been "mining" the soil by removing its organic content without replacing it, thereby compromising its future productivity for the sake of the enormous harvests of today. Soil science is in its infancy; it is a terrible illusion to think that we know how to keep soils healthy and productive, that we can take over the ancient expertise of those living organisms that create and maintain the soil habitat they need.

Feeding Ourselves Sustainably

Industrial agriculture that spurred the huge increase in productivity during the Green Revolution of the 1960s rests upon widespread use of highly selected strains of plants that require heavy machinery to plant, tend and harvest; chemical pesticides to eliminate competing weeds and insect pests; and artificial fertilizers to stimulate growth. A great deal of the value of food is the energy captured from the sun by photosynthesis and stored in sugars, fat and carbohydrates. We expend some of this energy by using our own muscles or the efforts of an animal, but for every unit of energy used by muscles, it is possible to harvest six to ten units of energy from the plant. Industrial agriculture exploits the energy in fossil fuels to synthesize pesticides, fix nitrogen for fertilizer and propel machines. For every unit of energy recovered from plants, industrial agriculture expends six to ten units in fossil fuels. In other words, the enormous productivity of modern farming results from converting fossil fuels to food with a *net loss* in the conversion. This cannot continue, since the resources that support this style of agriculture are non-renewable and in increasingly short supply.

The recent history of Cuba provides an example of how we can meet this crisis. As a satellite of the Soviet Bloc, Cuba flourished with a vast supply of Soviet oil and an assured premium market for sugar, its primary export. Heavily mechanized and dependent on chemicals, Cuban agriculture was as energy intensive as farming in any industrialized nation. In 1990, the Soviet Bloc collapsed with catastrophic consequences for Cuba—its source of oil and market for sugar disappeared overnight. The tiny nation didn't even have fuel to transport food from the countryside to the cities. Average urban caloric intake fell by 30 per cent.

Out of that crisis emerged an incredible success story—urban agriculture in which more than 80 per cent of the vegetables consumed in cities is grown in those cities. Yards and vacant lots were put into food production, while the state provided expertise and equipment to get farmers started. Today, more than ten thousand urban farms provide fresh, cheap, organic food to local neighbourhoods in Cuban cities while providing incomes and jobs to tens of thousands of people. The gardens achieve a productivity level comparable to yields from industrial agriculture. They do it by mimicking nature rather than bludgeoning it into submission with chemicals, artificial fertilizers and excessive irrigation. As in an ecosystem, a diversity of plants is grown to reduce targets for pests. Naturally occurring insecticides and fungicides are applied, and worms are used to process wastes and provide nutritious casts. This kind of small-scale agriculture requires a lot more physical exertion, but with society facing a massive obesity crisis, that too can be a plus.

There are many individuals and organizations, like those in

Cuba, that are working hard to give back to the soil, nurturing and restoring it, and honouring the reality that only the soil—a precious and endangered resource—can feed humanity. Miguel Altieri, an agroecologist, views agricultural land as an ecosystem in its own right. Altieri and his team use their understanding of the ecological principles to design and manage sustainable agriculture. By enhancing biodiversity, both within the soil and in the crops planted in the soil, his projects help farmers grow food year-round while reducing their reliance on chemicals and rebuilding the capacity of land to grow food. But beyond creating sustainable agriculture and ecological health, Altieri's projects strive to be culturally sensitive, socially just and economically viable. In his words, "sustainability is not possible without preserving the cultural diversity that nurtures local agricultures. Stable production can only take place within the context of a social organization that protects the integrity of natural resources and nurtures the harmonious interaction of humans, the agroecosystem, and the environment."

Although there continues to be a push toward ever more industrialization and globalization of the world's agriculture, to the detriment of worldwide food security, traditional livelihoods and the health of our land and water, many people like Altieri and his colleagues are pushing back. By working on a small, local scale and by challenging the myths and half-truths touted by corporate-controlled agriculture, they are rejecting a technological fix and instead are returning to the wisdom of the land and the people who have worked it for generations.

In developed countries, many people are speaking with their wallets. The market for organic and local food continues to grow. This increasing interest across the developed world in organic farming methods and pesticide-free produce suggests that the Earth's old wisdom is beginning to be heard again and that the soil is returning to its place at the centre of human life. It is our source of the very building blocks of life itself. Its productivity and health are a crucial link in the chain of life upon which we depend. Without exaggeration we can say the soil is the ground of our being; along with water and air it is the stuff that life helped to make, maintains and depends on absolutely.

Land then, is not merely soil; it is a fountain of energy flowing through a circuit of soils, plants and animals… An ethic to supplement and guide the economic relation to land presupposes the existence of some mental image of land as a biotic mechanism. We can be ethical only in relation to something we can see, feel, understand, love, or otherwise have faith in.

—ALDO LEOPOLD, *A Sand County Almanac*

5

The Divine
Fire

I am that supreme and fiery force that sends forth
all the sparks of life. Death hath no part of me, yet do I allot it,
wherefore I am girt about with wisdom as with wings.
I am that living and fiery essence of the divine substance that flows
in the beauty of the fields. I shine in the water, I burn
in the sun and the moon and the stars. Mine is the mysterious
force of the invisible wind... I am life.

—HILDEGARDE OF BINGEN, quoted in David Maclagan, *Creation Myths*

THERE IS NO ignoring the engine that drives the Earth and
all its life. Lift up your head on a clear morning and you will
see the sacred fire rising to the east. "We live," Wallace Stevens
says, "in an old chaos of the sun, / Or old dependency of day and
night," and so does every living thing, past, present and future.
Most human beings picture nothingness as being cold and dark;
since we first looked up into the sky we have been chanting our
devotion to the sun. In the first of the sacred books of Hinduism,
the Rig-Veda, written more than three thousand years ago, God
himself is created by heat:

In the beginning was darkness swathed in darkness;
All this was but unmanifested water.

Whatever was, the One, coming into being,
Hidden by the Void,
Was generated by the power of Heat.

That great fiat "Let there be light" set the ball rolling, started the flow of energy that made life possible and has echoed through the ages ever since.

Physicists define energy as the capacity to do work. They have learned that energy cannot be created out of nothing; it must be obtained from somewhere else. This insight has led to one of the most fundamental principles in science—the first law of thermodynamics, which states: "The total amount of energy in the universe remains constant. More energy cannot be created; existing energy cannot be destroyed. It can only be converted from one form to another."

So when a nail is driven by a hammer, the energy used to move the muscles of the body to deliver the blow comes from stores in the body, which in turn are recovered from food, which contains energy obtained from photons from the sun. When the nail is struck, energy is transferred to it and dissipated as heat—in the nail and in the wood and air surrounding it.

Energy stored in substances such as wood or gas is "high quality" because it can be readily obtained to do work. But when that energy is dissipated into water or air as heat, it becomes a low-quality form of energy. This insight leads to the second law of thermodynamics: "The spontaneous direction of energy flow is from high-quality to low-quality forms. With each conversion, some energy is randomly dispersed in a form that is not readily available to do work." This state of randomness or disorder is called entropy; we express the second law of thermodynamics in the vernacular by saying that everything tends toward disorder, or high entropy.

Without energy, life would not be possible. Life is the organic expression of energy. To move, to breathe, to see, to grow, to metabolize, energy is needed. Living things have a high degree of organization that requires much high-quality energy to keep them running. (Even when we are fast asleep, our bodies generate as much heat as a 100-watt light bulb.) That state of organization has increased through the number and complexity of living organisms on the planet over the past 3.8 billion years that life has existed. But if everything tends toward disorder, how has life been able to go on in spite of the second law? The reason is that energy from the sun is constantly flooding our planet, providing high-quality energy to compensate for the steady decay of energy. Without the addition of sunlight, life would soon run down. In the vacuum of outer space, the ambient temperature is a mere 3°K— that is, 3° above absolute zero, at which temperature all movement, including the movement of particles within atoms, would stop.

The Fire Within

We belong to a group of animals described as *homeothermic*, which means "constant temperature"; that is, our body temperature is maintained within a narrow range despite the fluctuations of temperature in our surroundings. Actually, it is only deep within the body that a constant temperature, about 37°C, is maintained; the temperature of our extremities and skin may vary by several degrees. In order to keep the core temperature constant, the amount of heat produced must equal the amount lost. Each one of us is like a house equipped with top-of-the-line air-conditioning and central-heating systems; a complex and sensitive thermostat continually adjusts the ambient temperature so that the house and its occupants are maintained in prime operating condition.

Metabolism—the process of burning fuels such as carbohydrates, fats and proteins—is our main source of heat. For a 70-kilogram man carrying out light work, the requirement of each type of fuel is as shown in Table 5.1.

‖‖

Table 5.1: Typical Fuel Requirement for a 70-Kilogram Man

MOLECULE	REQUIREMENT (G/DAY)	ENERGY (KJ/DAY)	% INTAKE
Fat	65	2500	25
Protein	70	1200	12
Carbohydrate	370	6300	63

Another source of heat is our skin. We may absorb heat from a radiant source such as the sun or a heat lamp or from direct contact with another object, such as a hot cup. We also absorb heat from the flow of hot water or hot air past our skin. In addition, we lose heat through our skin. We may lose up to a third of our heat by radiating it into our surroundings if they are cooler. If we sit on a cold metal chair, heat will flow from our body and we will feel cold. And we will lose heat if cold water or cold air flows past our skin. Scuba divers wearing wet suits derive some insulation from the rubber but are kept warm primarily because the flow of cold water past the skin is restricted.

Still another source of heat is muscular activity. When we are physically active, muscular activity can account for as much as 90 per cent of the heat produced in our bodies. When more heat is required because we are losing too much, then muscular activity is increased—either consciously when we become more active or involuntarily when we shiver. Infants have a larger ratio of surface area to internal volume than adults and so lose more heat through the skin. They have a deposit of "brown fat" in the shoulders and neck, which is metabolized to generate heat when stimulated by a drop in core temperature. As the surrounding temperature drops, thermoreceptors in our skin detect this drop and send signals to the hypothalamus of the brain, which sends signals commanding the smooth muscles of the blood vessels to contract and thus restrict blood flow. This phenomenon allows the extremities to cool while preserving heat for the central organs of the body by keeping the core temperature up. Our fingers can become cold because 99 per cent of the normal blood flow to them is stopped. Another set of smooth muscles is located at the base of hair follicles, an evolutionary remnant of our fur-covered ancestors. Stimulated by the hypothalamus, they contract and give us goose pimples.

When hypothermia sets in, these measures are no longer enough. In the early stages of hypothermia, when the core temperature is between 36° and 34°C, we may shiver and begin to breathe more quickly as blood is retained in our deeper body regions. We may feel dizzy and nauseated. At 33° to 32°C, shivering stops and metabolic heat production drops. When the core temperature reaches 31° to 30°C, we can no longer move deliberately, eye reflexes are inhibited, we lose consciousness, and the heart begins to beat irregularly. After the core temperature reaches 26° to 24°C, the ventricular muscles of the heart beat haphazardly and no longer pump blood. The house is closing down, and death soon follows.

When the temperature of air or water surrounding us is warmer than our skin, heat is absorbed by skin cells and spread to the rest of the body by blood. When the internal temperature of the body is higher than that of the skin, heat may be lost through the skin. If that is not enough to bring the body temperature down, our central nervous system receives signals that cause us to perspire through our sweat glands and cause water to diffuse through the skin. We have about 2.5 million sweat glands distributed on the skin of our body. As the warm liquid in sweat evaporates, it takes heat with it and we feel cooler. The amount we perspire is a reflection of both our internal temperature and the degree of ambient humidity.

Despite fluctuations in heat production, heat loss and heat uptake, our body temperature must be exquisitely balanced to maintain the optimum functioning under a wide range of surrounding temperatures. We have an average temperature of 37°C that fluctuates in a set way by about half a degree. During a fever or certain phases of the menstrual cycle, the deviation may remain for a prolonged period. There are heat-sensitive

Hypothermia as Healer

Despite the danger of a body becoming too cold, there are times when hypothermia can help heal. In some situations, medically induced hypothermia is a treatment option. Following a traumatic head injury, for instance, controlled cooling of the body core to about 33°C slows brain swelling, thus easing the pressure on the brain. Supercooling also minimizes the harm that the body's natural reaction to the injury can induce. Immediately after a trauma, the body begins to produce chemicals that can inflame the brain, damaging or even destroying its cells. Hypothermia slows this process, further protecting the brain from secondary damage. The body is cooled using ice packs, cold saline solution administered intravenously, or specially designed cooling blankets or vests. Cooling is also used during some heart surgeries and after cardiac arrest to curb the brain's appetite for oxygen and reduce the likelihood of brain damage.

In essence, this controlled use of hypothermia puts the body into a sort of hibernation where metabolism slows and the body uses less oxygen. This gives the body time to heal, or at least minimizes potential damage. When the danger to the patient has subsided, the body is slowly warmed back to 37°C, the temperature at which our metabolic engine burns most effectively. □

"thermoreceptors" located in the hypothalamus of the brain as well as additional sites in skin and the spinal cord. If the core temperature rises, the thermoreceptors detect the increase and send out messages to dilate blood vessels near the skin, increasing peripheral blood flow and thus dissipating heat brought from the core to the skin. Increased blood flow also decreases the opportunity for heat to be exchanged between arteries and veins. In addition, the central heat receptors stimulate sweating. If these measures are not enough to counter the increased heat, the core temperature rises a few degrees, enough to induce a potentially lethal state of hyperthermia.

We use heat as a defence against infectious disease; our body temperature rises in a fever, which is often lethal to the invading agent. A pyrogen, a protein from the invader, disturbs the thermoregulatory mechanism in the hypothalamus by setting the thermostat higher. Thus, relative to this setting, the body is cooler; as the fever sets in, chills and shivering accompany the rising temperature. As the fever subsides and the thermostat is returned to normal, the body has to be cooled; we begin to sweat and become flushed as more blood is sent to our skin to dissipate heat. Besides responding to a pyrogen carried on the surface of a bacterium, the liver and brain can generate proteins that will induce a fever. Like so many other weapons, this one is potentially dangerous to its user; high fevers can harm the body as they work to burn up the pathogens invading it.

Keeping the Internal Fires Burning

Older textbooks refer to animals as being either cold- or warm-blooded. This classification is a biological misnomer. Whereas mammals and birds do keep the temperature of their internal fire fairly steady, and so are warm-blooded, the internal temperatures of other animals—amphibians, reptiles, fish and invertebrates—vary with external temperatures. This means that on a warm day, a snake basking in the sun may be as "warm-blooded" as a mammal, but later, after the sun has set, its internal temperature will have dropped. Mammals and birds are endotherms (literally, "inner heat"), whereas all other animals are ectotherms ("outer heat").

Regardless of whether they are endo- or ectotherms, all animals have evolved behavioural and physiological strategies to regulate their temperature and keep their cells alive and bodies

The Promethean Bargain

The myth of Prometheus explores the ambiguities of fire power. Zeus had reserved the divine fire for the gods alone, but a cunning trickster in the Greek pantheon, Prometheus, stole it from the gods and brought it to men (there were no women in those cold suffering times, apparently). Later myths said he had actually created humans with that gift.

Such audacity could not go unpunished: Prometheus was chained to a mountainside where daily an eagle tore out his immortal liver. But humans suffered even more severely: Zeus could not remove the gift of fire, but he could craft another double-edged present. He created Pandora, the first woman, whose name means "all gifts," and sent her down to Earth carrying a sealed jar. Like fire, she was enchantingly beautiful, but she was also uncontrollably curious, unpredictable and deceptive. Inevitably she opened her vase; out swarmed Zeus's gifts—a horde of miseries to plague all humans for all time—disease, despair, rage, envy and old age were just a few of them.

This is the ancient Greek equivalent of the story of the Garden of Eden: human beings reached out recklessly, daringly, for knowledge and power and got more than they bargained for. And if we think of fire as the first true technology with which we began to change Earth, we can see the force of the story for our time. □

functioning. Mussels and tree frogs, for instance, produce natural antifreeze when temperatures drop to prevent potentially lethal ice crystals from forming in their body fluids. Amphibians cool as water evaporates off their moist skin, and reptiles bask in the sun to recharge their internal engines. Mammals and birds pant to cool down. As air passes across their warm, moist mouths, water evaporates, causing cooling. Penguins sit with their toes tipped up so that only their heels and tails are in contact with the ice. Seal lions and seals bask on sun-drenched rocks.

For endotherms, fur, feathers and fat are all part of the thermo-regulation tool kit. Air trapped between individual hairs or feathers creates a snug coat next to the skin. Coarser outer hairs or feathers repel wind, rain and snow. Thick layers of blubber help mammals survive in the ocean's depths and in the planet's highest and polar extremes.

Many animals, including marine mammals, large fish and birds, use an ingenious countercurrent heat exchange system to limit heat loss in cold air or water. Warm arterial blood pumped from the core of the body passes closely by veins coming from the extremities (e.g., feet and flippers) carrying cold blood back to the body core. Because of their close proximity, heat passes from the arteries to the veins. This process keeps the temperature of the body core constant and also cools the arterial blood slightly so that less heat is lost when the blood reaches the tip of the extremities.

Nature has equipped all animals with biological engineering to help keep the fires within burning. Only humans, however, have devised a way to carry, and use, an external source of heat.

A Fire Without

Our distant hominid ancestors were tree dwellers and, like our nearest current relatives, the Great Apes, were undoubtedly covered with fur. For reasons still in dispute, they descended from trees, stood upright and lost their fur. Scientists assume that they inhabited the tropical zone along the Rift Valley in Africa, where the severe fluctuations in temperature familiar in temperate regions were most likely not a problem. Nevertheless, the shifts in temperature from day to night and from the dry season to the rainy season and during severe storms must have been a challenge. Keeping warm by foraging for food, making clothing, constructing shelter and controlling fire must have exercised the observational and creative powers of those early humans. Perhaps those challenges helped to reinforce the selective pressures for greater cranial capacity.

The mastery of fire was perhaps the greatest milestone in our journey down from the trees and out into the wider world. With fire we escaped the limits of restricted habitats; carrying our warmth with us, we moved out of the tropics into the lands of winter—spreading out across Europe and Asia, moving eventually into areas as challenging as the Arctic tundra, the Himalayas and the Andes, and surviving well in Australia's central deserts with their enormous diurnal temperature range.

The Fire Deep Within

As our species moved from the tropical habitat where we originated to new territory where climate and temperature fluctuated to more extremes, we had to use our brains to find ways to conserve heat, inventing clothing and shelter and taming fire. But as biological beings, we already possessed a miraculous internal furnace.

The energy of a cellular substance refers to the usable energy that is released when it breaks down into simpler materials. This type of energy is known as chemical energy and is determined by burning a substance in oxygen and measuring the amount of energy released. That comes to 38.9 kilojoules per gram of fat, 17.2 kilojoules per gram of carbohydrates and 23 kilojoules per gram of protein.

Living cells are like minute stoves, extracting energy from fuel so that they can do work such as maintenance, growth or reproduction. Where does that energy come from? The earliest forms of life must have scavenged it from the chemical bonds of complex molecules. Take a lump of sugar and throw it into the fire; it will flare up and burn. Combustion breaks the chemical bonds between the atoms; they form new bonds with oxygen and break down into carbon dioxide, water and heat. The heat is the energy liberated by the broken bonds.

A cell can recover that energy of combustion but in a controlled way, by releasing the energy in successive steps and capturing that energy for use. In a cell, the equivalent of a match for starting a fire is a number of enzymes that break the bonds in a sugar molecule of glucose to create carbon dioxide, water and 15.7 kilojoules per gram of liberated energy.

In an atom, energy is absorbed by electrons, which then become more agitated. Electrons of an atom exist in different states of "excitation" that are related to the amount of energy they carry. If sufficiently energized or excited, an electron may release that energy by creating a chemical bond between two atoms. Cells have evolved a way of storing energy held by excited electrons in a special molecule called adenosine triphosphate, or ATP.

ATP is the source of energy that powers work in cells. ATP molecules act like chemical batteries, storing energy when it is not needed and releasing it as required. Figure 5.1 shows how the ATP cycle works. Since energy is stored in its chemical bonds, ATP energizes other molecules when it is instructed by an enzyme to transfer one of its three phosphate groups. This reaction releases a huge amount of energy that the organism can use. After the transfer, ATP is then ADP (adenosine diphosphate).

Turning ADP back into ATP takes energy from food or sunlight. Mitochondria (found in both plant and animal cells) extract the energy when this fuel is burned. Oxygen breaks the chemical

|||||||||||||||||||||||||||||||||||||||||||||||||||||||||||||||||||||||||||||||||||||||||||||||||||||

Figure 5.1: The ATP cycle.

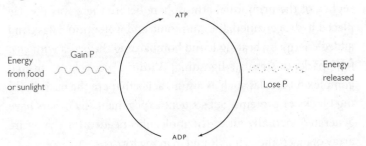

bonds and releases energy, which is fed back into the ATP cycle, reattaching a phosphate to ADP.

Organisms use ATP continuously, and this molecule moves through its cycle at an astonishing speed. In a working muscle cell, for example, 10 million molecules of ATP are consumed and regenerated each *second*.

Storing energy in ATP is like lifting a bucket of water onto a shelf. It takes energy to lift the bucket up, and when the water is poured from the bucket, some of that stored energy can be recovered from the force of the falling water. Or think about blowing air into a balloon. It takes energy, some of which can be recovered from the outrushing air when the balloon is punctured. In an analogous way, energy in excited chemical bonds can be recovered by breaking that bond. Cells possess hundreds of enzymes that are able to use that ATP-released energy to synthesize molecules, transport substances through membranes, move molecules such as muscle fibres and so on.

Let There Be Life (or the first zap)

Where did the energy come from to create life over hundreds of millions of years in those primeval oceans? The guess is that the first complex molecules were formed with energy contributed by lightning and by liberation of heat from molten magma, stream-ing down from volcanoes and up from deep sea vents. In a strik-ing experiment in the early 1950s, Stanley Miller set up a crude replica of the primordial atmosphere before there was life. He placed hydrogen, methane, ammonia and water into a flask and added energy by heating it and bombarding the gases with elec-tric sparks to simulate lightning. Within a week, he recovered complex molecules such as amino acids that are the basic build-ing blocks of proteins. Subsequent experiments by others have generated virtually all of the molecules needed for the entire array of macromolecules found in living forms.

There is much speculation about how the first cell arose and persisted. There must have been myriad bizarre life forms and test models as evolution kicked in to filter through them. In a way, it must have been like the early development of the automobile. Early cars went through many forms, with three, four or more wheels, some driven by steam, others by electricity or kerosene, some resembling buggies, others boxes and so on. But over time, certain underlying features became standard in all cars. So too with the first experiments in life.

Countless early models of cells must have been discarded as new mechanisms or structures appeared. But at some point not only did one cell survive and reproduce, it overwhelmed all the others to become the mother of all life forms on Earth. Today we believe that life cannot arise spontaneously, that life can only come from life. But once, at the very beginning, that first organism from which we are all descended was sparked into being, full of a life force that has so far persisted tenaciously for close to 4 billion years.

Scrounging for Energy

In that early period, life must have truly been brutish and short, a constant search for energy. Today, deep under the sea, heat from Earth's interior escapes through water columns superheated at high pressures to temperatures exceeding 250°C. Remarkably, bacteria exist in the boiling water column of those vents. Not only do they survive at these astonishingly high temperatures, they die of cold without them. These bacteria, which can recover energy from the hydrogen sulphide spewing forth, show us how some of the earliest species derived their energy. In a similar way, those primordial cells would have scrounged energy already existing in the chemical bonds within complex molecules. As long as complex molecules were abundant in the seas, living things could find the energy to grow, evolve and reproduce; bacteria flourished in the oceans for at least 2 billion years.

Today other organisms besides bacteria in deep sea vents also exploit chemical energy. Soil bacteria strip protons and electrons from ammonia molecules to obtain their energy while leaving nitrite and nitrate ions. Other organisms are able to exploit compounds of iron.

The pageant life has played out on evolution's stage creates an overwhelming impression of the extraordinary ability of living things to seize a chance and build on it. When some cells learned how to harness chemical energy for their own use, they became an opportunity for others. When they died, other organisms exploited their carcasses for the energy remaining in their molecules. Still other organisms became predators of the creatures who were those primary harvesters of energy; predators of the predators became yet another level of the food chain. Today a food chain of life forms feeding on energy and on each other is part of nature's balance.

Energy in the form of heat from the planet's interior alone does maintain populations of organisms, but they are restricted in number and distribution; the great explosion of complexity of animal and plant forms we take for granted could not have been supported from this source. The ultimate source of energy for most of life on Earth is the sun.

The Magnanimous Sun

Life on this planet is conferred by the fortuitous generosity of a rather undistinguished medium-sized star of a type called a yellow dwarf, one of the 400 billion stars in the Milky Way galaxy. From where we stand, it is a total marvel. About 99.8 per cent of all matter within our solar system is contained by the sun, 75 per cent of which is hydrogen and the rest helium. Nearly a third of a million times more massive than Earth, the sun has such an immense gravitational pull that the nuclei of its hydrogen atoms push through the repelling atomic forces and fuse. But the atoms formed by this nuclear fusion are unstable and so they "burn," emitting energy in the form of photons while being transmuted into helium.

With a diameter of 1.4 million kilometres, the sun has a surface temperature of 5777°K; at its core the temperature is a colossal 15 million degrees above absolute zero. Each second the sun burns 637 million tonnes of hydrogen to create 632 million tonnes of helium while releasing some 386 billion billion megawatts—the energy equivalent of 1 million 10-megaton hydrogen bombs. After 5 billion years aflame, the sun may be no more than middle-aged; it will probably burn for another 3 to 5 billion years.

Located 150 million kilometres from Earth, it radiates life-giving photons onto the planet. The solar wind of low-density charged particles—protons and electrons—streams perpetually toward Earth; when they hit atoms in the upper atmosphere they spark the release of light, and the skies dance with the shifting spectacle of the aurora borealis and aurora australis.

What a lucky chance it was that this particular planet coalesced at the right distance from our star. Earth's atmosphere, along with water vented from its interior, made it capable of creating the conditions for life. Water vapour and carbon dioxide were propitious molecules in the primeval atmosphere: they acted as greenhouse gases, keeping the sun's heat from escaping back into space.

Eventually life learned to eat sunlight through photosynthesis to stay alive—it was a giant step for evolution, a new level of metabolic innovation and biological adaptation. Between 2.5 billion and 700 million years ago, photosynthetic bacteria were the main occupants of the oceans, forming huge mats that remain today as fossilized structures called stromatolites. By 2.5 to 0.57 billion years ago, the oxygen by-product of the process had "polluted" the atmosphere; the air had become rich with this highly reactive, oxidizing agent to which life adapted.

We have the sun to thank for the oxygen in the atmosphere, which gave us aerobic metabolism, as well as another vital benefit—the ozone layer, which forms a shield against the ultraviolet component of sunlight, which damages DNA. Shielded from genetic damage, species could venture onto land.

In plants, photosynthesis occurs within defined organelles of cells called chloroplasts. Different pigments within the chloroplasts absorb photons of different wavelengths that we perceive as different visible colours of the rainbow. A dominant light-absorbing pigment is chlorophyll, which we see as green and which masks the other pigments in a leaf. In the autumn, when leaves stop producing pigments and chlorophyll breaks down, other pigments such as carotenoids are unmasked, appearing as red, orange and yellow and giving us the delight of fall colours. When a photon is captured by a pigment, its energy excites an electron, which is transferred to an acceptor molecule, which in turn passes it on to create ATP.

Chloroplasts are tiny photosynthetic factories—two thousand in a row would be thinner than a dime, yet each is bristling with thousands of clusters of two hundred to three hundred light-trapping pigment molecules. In these minute organelles, the productivity of life is fuelled by sunlight. And the chains of carbon atoms that are the sun's creation and life's signature have been a double-edged gift of residual energy to our generation.

A Legacy From the Past

Life has been playing out its drama on a vast planetary stage; and there has been enough rehearsal time to produce an extraordinary repertoire. You might say, looking at the diversity and proliferation of life today, that this production has run and run. Human beings have a limited perspective on time; we find it difficult to imagine how minute amounts of organic material in carcasses of bacteria, plants and animals can accumulate into massive deposits. Yet that is how life as we know it spread out across the planet. Over billions of years those bodies of dead organisms drifted down, piled up, drifted down, piled up on the ocean floor; more than 400 million years ago, as the planet's crust became geologically active, seas were drained and nutrient-rich sediments brought to the surface. Until this point the land had been barren of plants; now pioneers ventured into this new niche. Soon massive trees stretched into the sky, reaching for sunlight above the profusion of low shrubbery.

Between 280 and 360 million years ago, continents moved and sank, seas filled and drained some fifty times. Each time, some species disappeared while others took advantage of the changed environment. When seas drained, thick forests occupied swamps and lowlands; when these areas were again flooded, the organic matter in the forests was submerged in swampy water that had little oxygen to break down the plant carcasses. This organic matter, which was created by photosynthesis and metabolism, was made up of carbon atoms from the greenhouse gas carbon dioxide, which had been removed from the air during those processes. Micro-organisms decomposed the organic material, liberating oxygen and hydrogen and concentrating the carbon. Then the bacteria were killed by the acids liberated from the decaying plants. The partially decayed material is peat. As peat was buried under sediment, water and gases were squeezed out and the remnants were even richer in hydrocarbons. Initially, peat became a soft brown coal called lignite. As it became more deeply buried, lignite was transformed into a harder, darker

Tapping Chloroplasts for Fuel

The sun has made our planet habitable with its gift of pure energy. To be useful, however, the sun's energy must be converted from light energy to chemical energy. This process takes place in the chloroplasts of plants and algae. We use some of this energy when we eat food or burn wood or oil. Researcher Tasios Melis and his colleagues have discovered a way to harness nature's energy more directly, by flicking a "molecular switch" that will cause green algae to produce hydrogen gas instead of oxygen. In laboratory cultures, Melis found that in an absence of sulphur, the algal cells couldn't produce oxygen. This deprivation of sulphur disrupted the cells' normal respiration, and, to survive, the cells used an alternate metabolism, one which generated hydrogen, a natural gas that could—if enough volume could be produced—be an alternative, renewable and non-polluting fuel. Hydrogen fuel is used today, and, although it does not contribute harmful emissions when it is used to generate electricity, it is initially extracted from natural gas, a non-renewable resource, in a process that produces carbon dioxide as a waste product.

Although hydrogen gas siphoned from ponds of algae is not going to replace gas pumps any time soon, Melis's project is one of several investigating the possibilities of biohydrogen as an alternative to fossil fuels. □

material called bituminous coal. Then, as bitumen was subjected
to greater heat and pressure, it changed into anthracite.

Oil and gas are also made up of hydrocarbons from once-
living organisms. But whereas coal was formed from plants in
swamps, oil and gas came from marine plants and animals that
were buried in sediment that inhibited oxidation. Over millions
of years, the buried organisms were compressed and the organic
molecules underwent chemical changes to form petroleum and
natural gas. As they were compressed further, the oil and gas
moved upward through porous sedimentary rock until they
were trapped by an impermeable cap. Those accumulated reservoirs were a once-only gift of ancient life forms to an energy-
hungry industrial civilization.

Fossil fuels are the result of a long process in Earth's history,
a legacy of countless generations of life that flourished and died
with energy stored in the molecules of their bodies. It took hundreds of millions of years for this energy to accumulate and cook
into coal, oil and gas, and during all that time these substances
kept carbon out of circulation, helping to balance the proportion of greenhouse gases in the atmosphere. Now, in a flicker of
an eyelash, relatively speaking, the work of ages is being undone.

For most of its history, our species has burned animal fat,
dung, straw and wood as fuels. Coal has been used for just a
few centuries, and oil and gas are new fuels, in use only since
the Industrial Revolution. In this brief period we have suddenly
become dependent on fossil fuels on a global scale, depleting
easily accessible oil deposits at a rapid rate. Peak oil, the point at
which the global maximum rate of extraction has been reached,
is upon us. The race now is between fossil fuels and better alternatives. Cleaner energy sources and energy-efficient technologies
appear to be winning. Demand for oil, gas and coal is slowing
and expected price jumps have not materialized. Unfortunately,
demand is not yet slowing enough for us to even come close to

the Paris Agreement target of keeping the global average temperature from rising 1.5°C over pre-industrial levels. Global oil consumption declined somewhat during the COVID-19 pandemic, from close to 100 million barrels a day to about 91 million, but numbers are expected to climb again. As Colin Campbell, of the Association for the Study of Peak Oil and Gas, says, the more pressing issue to consider is the "downward slope that opens on the other side of peak production. Oil and gas dominate our lives, and their decline will change the world in radical and unpredictable ways." Despite affecting every aspect of our everyday lives, there is also the question of who will control oil resources and at what cost.

The first half of the oil age now closes. It lasted 150 years and saw the rapid expansion of industry, transport, trade, agriculture and financial capital, allowing population to expand six-fold. The second half now dawns, and will be marked by the decline of oil and all that depends on it...
—COLIN CAMPBELL, quoted in John Vidal, "The End of Oil Is Closer Than You Think"

As our population grows, we continue to squeeze what we can from the Earth's supply of fossil fuels, burning in an instant what took eons to create. Biologist Jeffrey Dukes calculated the amount of organic matter it took to create the fossil fuels most of humanity burns without a second thought. By his estimate, it takes about 90 tonnes of ancient plants to create 4 litres of gasoline. Clearly, fossil fuels are finite. They are a gift of sunlight and time. During the lifetime of our species, they will never again be created.

As well as depleting most oil reserves within a few generations, we are returning carbon dioxide to the atmosphere at a rate that exceeds the capacity of natural recycling mechanisms to remove it. For a century or more our use of energy has altered the amount of global atmospheric carbon dioxide. Although we can detect the changes in atmospheric composition, our ignorance of all the factors affecting climate and weather is so vast that we cannot predict all the consequences of these changes. But nature itself is giving signals that temperatures are changing as hurricane intensity increases, killer heat waves sweep Europe and Asia, ice sheets thin and glaciers retreat. Perhaps most compelling are changes in behaviour and ranges of flora and fauna: the latitude or altitude ranges of plant species are changing as they move to remain within the respective comfort zone of their species. Distribution of animal species such as abalone, for example, is changing with rising ocean temperatures; pine beetles are ravaging northern forests because they are no longer held in check by severe winters; and migratory birds arrive in the north weeks sooner and depart weeks later than usual.

Current models are remarkably consistent with the direction of the effects and in predicting the observed fluctuations in weather and temperature. Knowing that oil and gas will run out, that using them creates health and environmental problems and that there will be unpredictable climatological effects, we must clearly govern our use of energy within a program for ecological sustainability. Coal and peat deposits are vast, but they release even more greenhouse gases and are a greater problem.

It is clear where the chief responsibility for this crisis lies when Earth is viewed from space at night. As Malcolm Smith has described it:

> Most of sub-Saharan Africa, vast expanses of South America and central China are stark in their black vastness. North America, Western Europe and Japan, where a quarter of the world's population uses three quarters of the world's 10,000 million kilowatts of electricity, shine out as if we are hell-bent on advertising our profligacy.

A citizen of an advanced industrialized nation consumes in six months the energy that has to last the citizen of a developing country his entire life.
—MAURICE STRONG, quoted in the *Guardian*

We Have Met the Enemy and He Is Us

Glaciers are melting at an astounding pace, mountain pine beetles are eating their way through western Canada, sea levels are rising and we are shattering temperature records at an impressive rate—the ten warmest years up to 2020 have occurred since 2005, with 2020 and 2016 the warmest ever recorded. It takes hard work and a concerted effort to ignore the signs of global warming and climate change. Although there are still some holdouts, the vast majority of scientists agree that human activity is responsible for the warming of our planet.

The United Nations Intergovernmental Panel on Climate Change (IPCC) is one of the largest bodies ever assembled to assess a scientific issue. For its reports, thousands of scientists worldwide compile research on the state of the world's climate for policy-makers in 195 member countries. In its 2018 *Global Warming of 1.5°C* special report, the IPCC confirmed that "climate change is already affecting people, ecosystems and livelihoods all around the world" and that "limiting warming to 1.5°C is possible within the laws of chemistry and physics but would require unprecedented transitions in all aspects of society." Almost every country in the world signed on to the 2016 Paris Agreement to limit warming to 2°C, with an aspirational goal of 1.5°C. Seven countries signed but have yet to formally adopt its principles.

So, the scientific evidence is there. There is no doubt that global warming is happening and that human activity has propelled it forward at breakneck speed. It's well past time to act, yet the response of governments is mixed at best. The Paris Agreement was an important milestone, but it was weakened when former president Donald Trump pulled the U.S. out in 2017. Although President Joe Biden reversed that decision, it was a setback when the world could least afford it. Most countries, including Canada, are not on track to meet their greenhouse gas emissions reduction targets, and Canada's federal and provincial governments continue to support fossil fuel development. Countries like Canada that contribute an oversized share of the problem need to do more. Canada, for instance, makes up less than 0.5 per cent of the world's population yet is one of the world's largest per capita emitters of greenhouse gases.

Fortunately, many individuals, local governments and corporations, too, are not waiting for federal governments to lead. The city of Calgary, for instance, is using wind power to run its rapid transit system, and the province of Quebec has a climate plan that will help it achieve carbon neutrality by 2050 and adapt to unavoidable climate change consequences. Many American states and cities are taking steps to reduce energy costs and reduce greenhouse gas emissions. Even corporations are recognizing that reducing fossil fuel use makes good business sense. Although met with some skepticism, oil companies like BP have pledged to tackle emissions from their operations and products, with BP aiming for net-zero carbon emissions by 2050.

As scientist and writer Amory Lovins notes, "global warming is cheaper to fix than to ignore. Because saving energy is profitable, efficient use is gaining traction in the marketplace... The climate problem was created by millions of bad decisions over decades, but climate stability can be restored by millions of sensible choices."

Of course, there will be economic implications to resolving the climate crisis—and they'll continue to increase the more we delay taking necessary action. In 2006, former World Bank chief economist Nicholas Stern calculated in his report *The Economics of Climate Change* that it would cost about 1 per cent of the world's gross domestic product (GDP) to adequately deal with the problem of global heating but that not acting would cost trillions and destroy about 20 per cent of the global economy, plunging the world into an unprecedented economic depression.

More recently, a 2021 report from management consulting firm McKinsey & Company calculated that reaching net-zero global emissions by 2050 would require global spending on physical assets of about $275 trillion between 2021 and 2050, or about 7.5 per cent of GDP annually on average, with the biggest increase between 2026 and 2030. But the report points out that "the costs and dislocations that would arise from a more disorderly transition would likely be far greater, and the transition would prevent the further buildup of physical risks." As well, the report argues, "it is important not to view the transition as only onerous; the required economic transformation will not only create immediate economic opportunities but also open up the prospect of a fundamentally transformed global economy with lower energy costs, and numerous other benefits—for example, improved health outcomes and enhanced conservation of natural capital."

The Intergovernmental Panel on Climate Change's 2021–22 Sixth Assessment Report found that addressing the climate crisis wouldn't cost much more than what we're spending now to keep burning fossil fuels. (The International Monetary Fund says global subsidies alone for coal, oil and gas amount to US$11 million every minute—almost $6 trillion in 2020.) "Without taking into account the economic benefits of reduced adaptation costs or avoided climate impacts, global Gross Domestic Product (GDP) would be just a few percentage points lower in 2050 if we take the actions necessary to limit warming to 2°C (3.6°F) or below, compared to maintaining current policies," IPCC Working Group III co-chair Priyadarshi Shukla said.

Do we really have a choice?

Playing With Fire

All the gods of all our stories know that fire is a double-edged sword; what warms may burn, what gives power may also consume, what gives life may take it away just as easily. Our relationship with fossil fuels is just the latest evidence of this difficult, dangerous truth. Our use of energy in the industrialized world has given us comfort, economic security, mobility, food and the power to change Earth to suit ourselves. It has also given us a Pandora's box of associated miseries: air pollution, soil erosion and environmental destruction. Fossil fuels have provided cheap, portable energy to fuel vehicles and to manufacture machinery that has brought us that deadly affliction of overconsumption, which is clearing the world's forests, emptying its oceans, devastating its waterways, obliterating its non-human life. So how can we contain the power we have snatched so recklessly?

The mix of life, balanced yet constantly changing over time, teaches us the ground rules: what species do is local and small-scale and introduces little that is novel. In nature, the dung beetle lays its eggs in animal droppings to exploit the remnant food value of the dung. Plants bleached of chlorophyll survive by parasitizing photosynthetic green plant carcasses, only to become food for insects and other animals. There are cycles within cycles. In biological systems, the passage of energy and materials forms loops that are completely circular, and so there is no end product to be dumped into soil, air or water, the other elements we are pledged to protect.

Human beings have broken those loops, creating linear use of energy and matter that go from raw resources to heat and materials that are discarded or lost. Often there are unanticipated consequences to the buildup of these wastes. Intrinsic to our myths of power are warnings: technologies have unexpected side effects, and the bigger the technology, the more intractable the consequences. When Pandora opened her box, all the plagues that torment humanity scattered to the far ends of the Earth. But one thing was left behind: tucked into a corner was the welcome figure of Hope. And hope remains that we can achieve sustainable levels of energy consumption by making existing systems more efficient and by using alternative sources of energy: the sun, the wind, the tides and the deep, abiding heat of the Earth.

David Pimentel outlined an economy based on the sustainable use of energy, land, water and biodiversity while achieving a relatively high standard of living, but he noted that steps on a heroic scale must be taken immediately with a view to reducing both use of fossil fuels and population. Ninety million hectares of land (equal to the combined areas of the states of Texas and Idaho) could be used to collect solar energy without disrupting agricultural and forest productivity. By conserving energy, per capita consumption of oil could be cut in half to 5000 litres of oil equivalents. By conserving soil and water, reducing air pollution and massively recycling, a conserver society could be achieved in the United States in which

> the optimum population would be targeted at about 200 million... Then it would be possible for Americans to continue to enjoy their relatively high standard of living... Worldwide, resolving the population-resource equation will be more difficult than in the United States.

The global population level could reach 10 billion before the middle of this century. These projections, still relevant today, all suggest the need for a massive effort to conserve soil and to recover enough food for each person on 0.5 hectares of land. These measures will have to be accompanied by rapid stabilization and then reduction in population. If these goals are achieved, it would be possible to sustain a global population of approximately 3 billion humans. With a self-sustaining renewable energy system... providing each person with 5,000 liters of oil equivalents per year (one half of America's current consumption/yr but an increase for most people in the world), a population of 1 to 2 billion could be supported living in relative prosperity.

Although Earth's human population is almost 8 billion and growing, Pimentel's vision is still one of hope. It talks about focused effort, about conserving energy and sharing it out fairly; above all it proposes a new beginning. Fossil fuels now suffuse every aspect of our lives, used in our cars, furnaces, energy-dependent manufacturing, farming and so on. But we became dependent on this source of energy very recently. Now that we understand the repercussions of the finite nature of oil and gas and the buildup of greenhouse gases when we use too much, we can turn our creative energies to finding alternatives, especially by harvesting the energy flooding onto Earth from the sun. There is a great deal of opportunity. It will take time to wean ourselves from our current patterns of energy use, and we can extend that time by becoming much more efficient, stretching our reserves and reducing our contribution of effluent gases and our wastes. Automobile technology is changing rapidly in response to regulation and innovation. Zero-emission, hybrid and autonomous vehicles are helping to reduce some of the ecological impacts of transportation. This could buy time for the design and construction of living spaces for most of humanity that eliminate the need for cars altogether. Greater efficiency in manufacturing processes can reduce energy and materials use by a factor of four, while conservation through reduced consumption can solve the ecological problems and increase equity. The potential is there. What is needed is the will.

6

Protected by
Our Kin

There is grandeur in this view of life...
whilst this planet has gone cycling on according to the fixed law
of gravity, from so simple a beginning endless forms most
beautiful and most wonderful have been, and are being, evolved.
—CHARLES DARWIN, *On the Origin of Species*

... the multifarious forms of life envelop our planet and,
over aeons, gradually but profoundly change its surface. In a sense, life
and Earth become a unity, each working changes on the other.
—LYNN MARGULIS, *Five Kingdoms*

E VERY CHILD WHO has marvelled at the growth of a plant
from a seed, observed the transformation of a frog's egg
into a tadpole or witnessed the emergence of a butterfly from its
cocoon understands in the most profound way that life is a miracle.
Science cannot penetrate life's deepest mystery; music and poetry
attempt to express it; every mother and father feels it to the core.

To the centre of the world you have taken me and showed the goodness
and beauty and strangeness of the greening Earth, the only mother.
—BLACK ELK, quoted in T.C. McLuhan, *Touch the Earth*

Early thinkers recognized the four elements necessary for life—air, water, earth and fire. But they did not know that the collective effect of living things themselves had played a vital hand in shaping and maintaining those elements. Life is not a passive recipient of these elemental gifts but an active participant in creating and replenishing them.

A thought exercise is useful to illustrate the critical role that all life plays in providing what Indigenous people refer to as the four sacred elements: earth, air, fire and water. Imagine that scientists have created a time machine that takes us back 4 billion years before life arose on this planet. If we rush out of the time capsule to investigate this sterile world, we'd be dead in minutes because the prebiotic atmosphere, although rich in water vapour and carbon dioxide, lacked oxygen. It was only after life discovered photosynthesis that oxygen was released as a by-product of the capture of sunlight. This process transformed the atmosphere over millions of years, producing the air that animals like us depend upon.

Suppose we anticipated these inhospitable conditions and have stored tanks of air that we can strap on before exploring the Earth. After a few hours in the warmth (water and carbon dioxide are greenhouse gases), we would get thirsty, but any water would be questionable for drinking because there are no plant roots, soil fungi or other micro-organisms to filter out heavy metals and other potentially dangerous leachates from rock. We would get hungry, but, of course, since every bit of the food we eat was once alive, there would be nothing to eat. Even if, in addition to a supply of food, we brought some seeds to grow fresh vegetables, we would find no soil in which to grow anything because soil is created when living organisms die and their carcasses mix with the matrix of clay, sand and gravel.

And suppose at the end of the day on this lifeless planet, we feel homesick and decide to light a campfire for comfort. There would be no fuel to burn because every bit of our fuel—wood, dung, peat, coal, oil, gas—is formed by life. Furthermore, even if we had brought fuel, we couldn't burn it because without oxygen, no flame could ignite. This incredible journey through time reveals that the web of all life keeps the planet hospitable.

Life maintains its unique handiwork by means of its extraordinary power to diversify—to adapt to opportunities as they present themselves and to create new opportunities in the process. No single species is indispensable, but the totality of all life forms maintains the fecundity of Earth. Thus, the diverse array of life itself may be regarded as another of the fundamental elements that support all living things. Biodiversity must take its place beside air, water, earth and fire, the ancient creators of the planet's fertility and abundance.

By having the sun rise every day, the earth and the sky are renewing their commitment to work together. And we in turn honour that relationship and directly benefit from it.

—AIMÉE CRAFT, *Treaty Words*

Life and Death: Conjoined Twins

Life and death are a balanced pair. It is a strange irony that death has been a critical instrument in the persistence of life. Humanity's age-old dream of eternal life, if ever realized, would lock any species into an evolutionary straitjacket, eliminating the flexibility required to adapt to the planet's ever-changing conditions. By allowing adaptive change to arise in successive generations, individual mortality enables species to survive over long periods of time.

In the end, however, the species proves as mortal as the individual. Over the sweep of evolutionary time, it is estimated that 30 billion species have existed since multicellular organisms arose in the explosion of life in the Cambrian era, 550 million years ago.

On average, scientists believe, a species exists for some 4 million years before giving way to other life forms. It is estimated that there may be about 30 million species on Earth today—that means 99.9 per cent of all species that have ever lived are now extinct. But all forms of life on the planet today have their beginning in one cell that arose in the oceans perhaps as long as 3.8 billion years ago, and from the perspective of the vital force imbued in that first cell, life has been astonishingly persistent and resilient.

The Interconnectedness of All Life

The forest is one big thing—it has people, animals and plants. There is no point in saving the animals if the forest is burned down. There is no point in saving the forest if the animals and people are driven away. Those trying to save the animals cannot win if the people trying to save the forest lose.

    —BEPKOROROTI, quoted in "Amazonian Oxfam's Work
      in the Amazon Basin"

No species exists in isolation from all others. In fact, today's estimated 30 million species are all connected through the intersection of their life cycles—plants depend on specific insect species to pollinate them, fish move through the vast expanses of the oceans feeding and being fed upon by other species, and birds migrate halfway around the world to raise their young on the brief explosion of insect populations in the Arctic. Together, all species make up one immense web of interconnections that binds all beings to each other and to the physical components of the planet. The disappearance of a species tears the web a little, but that web is highly elastic. When one strand is rent the whole network changes configuration, but so long as there are many remaining strands to hold it together, it retains its integrity.

We have to feel the heartbeats of the trees, because trees are living beings like us.

—SUNDERLAL BAHUGUNA, quoted in E. Goldsmith et al.,

Imperiled Planet

All life ultimately depends on energy from the sun, which is exploited by plants and micro-organisms through photosynthesis (as we have seen, only a very small number of microbial species, which are said to be chemosynthetic, can oxidize inorganic substances such as nitrogen and sulphur to obtain energy or can utilize energy coming from the core of the planet). The primary consumers of the photosynthetic and chemosynthetic organisms are herbivores as varied as grasshoppers, deer and krill, which in turn provide sustenance for primary carnivores such as spiders, wolves and small squids. Secondary carnivores such as toothed whales, eagles and humans feed on primary carnivores and are furthest away from the original exploiters of energy. Eventually, all parts of the network will be reprocessed by decomposing organisms and returned to the Earth (Figure 6.1).

The Invisible World

It is humbling to realize how restricted our perception is compared with other creatures on this planet. Our view of the world is created by the degree of sensitivity of our sensory organs. We are aware of how limited this can be each time we watch the peregrinations of a dog as it runs from hydrant to tree, breathing in a world of impressions, the chemical signatures left by other animals that indicate their age, sex and species, as well as how long ago they were there. Insects can respond to a single molecule of pheromone floating in the air. Other animals, from black-tipped sharks to fiddler crabs, sense changes in barometric pressure and can therefore anticipate changes in weather long before humans

|||||||||||||||||||||||||||||||||||||||||||||||||||||||||||||||||||||||||||||||||||||||||||||||||||||||||||||||||||

Figure 6.1: A food chain in a temperate ecosystem.

Adapted from *Science Desk Reference* (New York: Macmillan, 1995), p. 463.

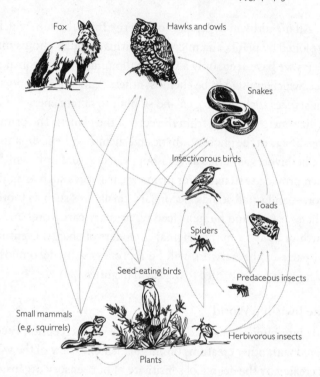

can. Our ears lack the ability to detect the high-pitched sounds that help bats manoeuvre, capture prey and avoid predators. We are deaf to the low-pitched frequencies that are the songs of marine leviathans echoing through oceans halfway around the world. The seismic communication of elephants—vibrations received through the feet and nerve-riddled, ultra-sensitive tip of the trunk—pass by us undetected. Our vision is limited to wavelengths of light that our sense organs can detect in the range from red to purple. We can't see infrared as the rattlesnake can or the ultraviolet light that guides insects to specific flowers.

As air-breathing animals, we are ignorant of the vast range of diverse marine and freshwater ecosystems and the plants and animals that have adapted so wonderfully to them. Being held to Earth's surface by gravity, few of us have seen the planet as a soaring bird has or as members of communities dwelling in forest canopies have. Nor do we have the subterranean perspective of the burrowing animals, plants and micro-organisms that spend most of their lives beneath our feet. As animals of the day, we are insensitive to the interplay of creatures that are active at night.

Our light receptors cannot resolve objects in the size range of single cells, and so we are blind to the vast numbers and variety of microscopic life in a single drop of pond or ocean water or a pinch of soil. Of course we have compensated for our physical shortcomings by creating technologies that extend our sensory range. We detect the symphony of inaudible sounds through machinery that can make their patterns visible or audible. We can detect extremely low concentrations of molecules—drugs, explosives, DNA—in the air or adhering to objects.

But it is microscopy that has opened a whole new world to us. What a wondrous shock it must have been to the pioneers who first saw the cosmos of bizarre forms in staggering abundance and variety revealed by magnifying lenses. These miniature organisms were the only life forms for most of the time that Earth has been animated, and even today they have a biomass equivalent to or greater than that of all of the ancient forests, great herds of mammals, vast flocks of birds, enormous schools of fish and countless insects taken together. (For some sense of scale, in her book *Garden of Microbial Delights*, Lynn Margulis tells us there are a hundred thousand microbes per square *centimetre* of human skin!) This vast universe of life, invisible to our species, has carried on as the dominant organisms on the planet for billions of years. As we marvel at the large creatures—ancient trees, birds, mammals—we owe our very existence to the teeming universe of microscopic lives.

We live now in the "Age of Bacteria." Our planet has always been in the "Age of Bacteria," ever since the first fossils—bacteria, of course—were entombed in rocks more than 3 billion years ago.

—STEPHEN JAY GOULD, "Planet of the Bacteria"

Tinkering With Life

One of modern biology's great insights has been the recognition that DNA is the blueprint of life, dictating the physical makeup of all multicellular organisms. By elucidating its molecule structure as a double helix, James Watson and Francis Crick, as well as Rosalind Franklin, began a revolution that now allows scientists to create organisms virtually at will. Today, scientists can isolate, purify, sequence and synthesize specific genes and then transfer them between unrelated species. This ability has led to an explosive growth in biotechnology, wherein spectacular new organisms are created by gene transfer: strawberries resistant to frost because of an implanted fish gene that produces antifreeze; rice rich in blindness-preventing vitamin A; bananas implanted with genes allowing them to produce antibiotics. The list is restricted only by one's imagination. The notion of creating designer organisms for human benefit is irresistible.

Biotechnology is trumpeted as a means to eliminate starvation and suffering by increasing yields for a growing human population, creating crops resistant to pests and generating new drugs. Yet the risks of genetically engineered organisms or their products, like the risks of DDT or CFCs when they were first introduced, are largely unknown, because our basic knowledge about how cells, organisms and ecosystems work is too limited to allow us to anticipate the repercussions of manipulating these organisms' genes. The terrible error in biotechnology is thinking that genes exist and function in isolation. A gene is part of a greater, integrated whole—the genome—which has been selected and honed to turn off and on whole suites of genes in proper sequence and timing from fertilization to maturity, a network of gene relationships and connections we are just beginning to tease apart and reveal. A gene transferred from one species into another finds itself in a totally alien context leaving us little ability to anticipate consequences, much like removing Mick Jagger from the Rolling Stones and inserting him into the New York Philharmónic orchestra and asking him to make music. Sounds will emerge, but whether they will be music is unknown.

It is the *context* that makes a gene relevant. As the late Richard Strohman, a biochemist and former head of molecular and cell biology at the University of California, Berkeley, said:

> When you insert a single gene into a plant or animal, the technology will work... you'll get the desired characteristics. But you will also... have produced changes in the cell or the organism as a whole that are unpredictable... Genes exist in networks, interactive networks which have a logic of their own... And the fact that the industry folks don't deal with these networks is what makes their science incomplete and dangerous... We are in a crisis position where we know the weakness of the genetic concept, but we don't know how to incorporate it into a new, more complete understanding. □

Nature Is Cyclical

Natural systems are deeply entwined—and they are circular, one species' waste becoming another's raw materials or opportunity so that nothing goes to waste (Figure 6.2). The cyclic linking of different species is illustrated by the exquisite life cycle of the five species of Pacific salmon, which are renowned for their incredible abundance. Even though fewer than one in ten thousand fertilized eggs may reach adulthood, the survivors return from the ocean to their natal streams at maturity by the tens of millions. From the moment a salmon begins life at fertilization, it runs a gauntlet of predators—trout, ravens and fungi in fresh water; killer whales, eagles and seals when it migrates to the oceans. Even in death salmon provide nourishment: their carcasses are food for bacteria and fungi, which feed microscopic invertebrates, which eventually nourish the emerging fry that are the salmon's own offspring. Birds and mammals, bellies swollen with their bonanza of salmon carcasses, spread nutrients from the salmon across the forest floor in their droppings. To human predators, the salmon life cycle may seem "excessive" or "wasteful," but in the cycle of living things, nothing goes to waste.

Early in the history of life, Nature began to shape new species to fit into habitats already occupied by other species. Never since the Archaean Period has a living thing evolved alone. Whole communities have evolved as if they were one great organism. Thus all evolution is coevolution and the biosphere is now a confederation of dependencies.
—VICTOR B. SCHEFFER, *Spires of Form*

Human beings depend on Earth and its life forms for every aspect of their survival and life. It is impossible to draw lines that delineate separate categories of air, water, soil and life. You and I don't end at our fingertips or skin—we are connected through

||||||||||||||||||||||||||||||||||||||||||||||||||||||||||||||||||||||||||||||||||||||||||||||||||||||||

Figure 6.2: Groups of organisms classified by food consumption.

Adapted from Cecie Starr and Ralph Taggart, *Biology: The Unity and Diversity of Life*, 6th ed. (Belmont, CA: Wadsworth, 1992), fig. 40.8.

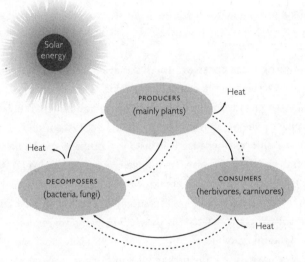

air, water and soil; we are animated by the same energy from the same source in the sky above. We are quite literally air, water, soil, energy and other living creatures.

Why Biodiversity Is Important

From the earliest times we humans have used our massive brains to exploit the variety of species surrounding us. We learned which plants were edible and how to catch animals that were faster or stronger than we were. We learned how to use the natural defences of animals and plants, tipping arrows with poison, stunning fish in rivers. The medicinal properties of other species healed our ills, their beauty decorated our bodies, and their skins protected ours, as clothing and shelter. The diversity of living things in different ecosystems is demonstrated by the range of uses we have found for other creatures as we spread and settled across the world.

Salmon Forests

Along the west coast of North America, pinched between the Pacific Ocean and the coastal mountains, is a temperate rainforest that stretches from California to Alaska and boasts the greatest biomass (weight of living things) of any ecosystem on the planet. It is a rainforest because it rains a lot, but one of the mysteries of this ecosystem is how such huge trees—red and yellow cedar, Douglas fir, Sitka spruce, hemlock and balsam—can flourish when essential nitrogen is in limited supply because it is washed from the soil. The answer to this puzzle illustrates the exquisite interconnectedness of life.

We have long known that salmon born in coastal rivers and streams need the forest to keep the waters cool, to retain the soil (which, in turn, prevents erosion) and to provide feed for baby fish, because when a watershed is clearcut, salmon populations plummet or disappear. But now we are learning that the forest needs the fish, too.

Almost all of the nitrogen in terrestrial ecosystems is the isotope nitrogen-14 (^{14}N), but in the oceans, there is a relatively high concentration of the heavier isotope nitrogen-15 (^{15}N). When the salmon go to sea, they consume ^{15}N-laden prey and accumulate the isotope in their tissues. Upon reaching maturity and migrating back to their natal streams, the salmon's protoplasm is laden with ^{15}N. Eagles, ravens, wolves, bears and dozens of other organisms feed on the carcasses of spawned-out salmon and then distribute the marine nitrogen throughout the forest in their feces. During the spawning season, bears may consume up to six hundred fish each.

They usually carry a captured salmon up to 150 metres away from the river before eating part of it and then returning to catch another. The remains of the carcass are then eaten by salamanders, beetles, birds and other creatures, including flies that hatch as maggots. The ^{15}N-filled maggots mature and fall onto the forest litter, where they pupate over winter to emerge the next spring as flies in time to feed birds migrating from South America on their way to the Arctic. Salmon that die in the river sink to the bottom and are soon covered in a thick blanket of fungi and bacteria, which in turn feed insects and other invertebrates. So when the fry emerge from the spawning gravel four months later, the waters are filled with a banquet of ^{15}N-laden food that fed on the carcasses of their parents.

And now the mystery of the huge trees in coastal rainforests is solved. Salmon represent the single largest pulse of nitrogen fertilizer spread by other creatures that the trees get all year. That record can be deduced by measuring the amount of ^{15}N in tree rings and correlating those data with the size of the annual runs.

Humans, with our political, economic and social priorities, assign various facets of the salmon's vast reach to different ministerial departments. Departments responsible for commercial, sport and Indigenous food fisheries handle the salmon themselves; the department of forestry handles the trees; environment, the whales, eagles and bears; agriculture and energy, the rivers; mining, the mountains and rock; and so on. We fail to account for the interconnectedness of ocean, forest and northern and southern hemispheres, thereby fragmenting the integrity of this system and guaranteeing that we will never manage it sustainably. ☐

When we domesticated animals and plants, only ten thousand to twelve thousand years ago, human life changed forever, vaulting to another level in the evolution of culture. All the domesticated animals and plants that human beings depend on today were once wild, and we continue to need the genetic diversity that exists in wild populations—that diversity is still life's major defence against changing conditions. For this reason alone humanity has an absolute need to protect biological diversity: it is a matter of sheer self-interest. Biodiversity has its own worth regardless of how it serves people. As French philosopher Catherine Larrère says, "All living organisms, through their existence and their use of complex, non-mechanical strategies to survive and reproduce, have their own value. Beyond that, biological diversity itself, because it is the product of evolution and also the condition for its continuation, has its own intrinsic value..."

Another compelling argument for protecting biodiversity is the unfortunate fact that we know next to nothing about most species on Earth. We know there is a web of life, but every time we study a small section of the web we discover what seems to be an infinity of interconnections. The more we learn, the more we realize how much else there is to learn about the way life acts and interacts to survive.

The Molecular Blueprint

Since the beginning, Native Peoples lived a life of being in harmony with all that surrounds us. It is a belief that all humankind are related to each other. Each has a purpose, spirit and sacredness. It is an understanding with the Great Spirit or Creator that we will follow these ways. And in this understanding we believe we are related to all other living species.

—DENNIS BANKS, quoted in Native American Netroots

By studying DNA, molecular biologists have verified that all living organisms are genetically related. The central novelty of the movie *Jurassic Park* was the discovery of an ancient mosquito preserved in amber whose gut carried intact pieces of dinosaur DNA. Were it a true story, the even more remarkable fact would have been that both the mosquito's DNA and the dinosaur's DNA could be shown to carry segments identical to genes found in every one of us. Through our evolutionary history, we are related to all other beings present and past—they are our genetic kin. When we see other species as our relatives rather than as resources or commodities, we will have to treat them with greater care and respect. In the words of Black Elk:

> It is the story of all life that is holy and is good to tell, and of us two-leggeds sharing in it with the four-leggeds and the wings of the air and all green things; for these are children of one mother and their father is one Spirit.

Indeed, all life forms are our relations. Whereas it may not be too hard to grasp that humans and apes share about 98 per cent of their genes, it may be more of a stretch to realize that humans share about 85 per cent of their genes with mice. What's more, we carry hundreds of genes that are similar, and in many cases identical, to genes found in fruit flies, roundworms, yeast and even bacteria.

The evolutionary unity of humans with all other organisms is the cardinal message of Darwin's revolution for nature's most arrogant species.
—STEPHEN JAY GOULD, *The Mismeasure of Man*

In Praise of Genetic Diversity

How does life achieve its extraordinary resilience? In the early 1960s, when new biochemical techniques were developed, scientists began to analyze the products of specific genes carried by individuals of a species. To their great surprise, the biologists discovered a large number of hitherto undetected gene variants, or different forms of the same gene, within a species. Geneticists refer to this diversity as genetic polymorphism; it seems to be the means by which a species responds to changing environmental circumstances. Most gene variants apparently have little or no effect on the way the product they specify functions, so they are referred to as neutral differences, neither beneficial nor detrimental in a given environment.

But neutrality is temporary and relative. When conditions in the surrounding environment change—in acidity, salinity or temperature, for example—then different forms of one gene can specify products having quite dissimilar functional activities or efficiencies. In humans, a classic example is a gene variant or mutation called sickle cell that affects hemoglobin in the blood. When people carry two copies of the mutant gene (inheriting one from each parent), they suffer from a condition known as sickle-cell anemia, which is extremely painful and often lethal. Those who carry one copy of the sickle gene and a normal gene are normal, except in places where malaria is rampant. In such places these people have a greater resistance to the parasite than people who carry two normal genes.

The Importance of Variation

In my own work with the fruit fly, *Drosophila melanogaster*, I was able to recover chemically induced mutations that are invisible (that is, not expressed) under certain conditions but produce an abnormality when grown under a different environmental regime. The mutations I studied were influenced by temperature—at one temperature, the flies were completely normal, yet a shift of as little as 5° or 6°C would result in a variety of mutant expressions. I discovered such environmentally determined expression of genes causing everything from visible abnormalities in wings, eyes or legs to reversible paralysis or death. When global weather patterns and average temperatures fluctuate as a result of climate change, those species with genes that enable individuals to function properly or better at the new temperatures will be the survivors who will carry on.

... in the great majority of species, somewhere between 10 and 50 percent of genes are polymorphic. A typical figure is roughly 25 percent.
—EDWARD O. WILSON, *The Diversity of Life*

Genetic polymorphism is crucial to a species' survival. When a species such as the whooping crane or Siberian tiger is reduced to a handful of survivors, its long-term future is in doubt because the range of its genetic variability has been radically diminished. Thus, it has fewer options for adapting to changes in the environment. Furthermore, in a small population, there is a greater likelihood that recessive genes that are lethal or that threaten viability when two copies are present will be exposed. A diverse mixture of gene variants is a fundamental characteristic of a vibrant, healthy species, a reflection of its successful evolutionary history and continued potential to adapt to unpredictable change.

Population geneticists believe that the most successful species (where success is defined by long-term survival) are found in many isolated pockets or islands that are connected by "bridges" across which a constant trickle of individuals passes. Thus, each isolated community can evolve a set of genes adapted to its local habitat, while the occasional migrant becomes a means of introducing "new blood"—different genes with a new potential to respond to change.

In recent times, large-scale industrial agriculture has taught us an expensive lesson—reducing genetic diversity by the widespread use of a single selected strain of a crop, known as monoculture, is extremely risky because it makes a species vulnerable to change. In 1970, approximately 80 per cent of the 26.8 million hectares planted in corn in the United States carried a genetic factor for male sterility. But that trait, so useful to seed companies, was its Achilles heel, rendering the strains vulnerable to a specific parasite. Within three months, a devastating southern corn blight had swept across the continent, affecting virtually all fields. Overall losses were 15 per cent, but many farms lost 80 to 100 per cent of their corn that year for a total cost of $1 billion.

Monoculture counters life's evolutionary strategy. In fish hatcheries, the broad genetic polymorphism of wild stocks of fish such as salmon is displaced by large numbers of hatchery-reared fingerlings grown from eggs and sperm taken from a few fish selected according to their size. Again, this type of selection reduces genetic diversity, and decreased diversity is one of the causes of catastrophic declines in salmon returns. Foresters have belatedly recognized that tree plantations of fast-growing strains of commercially valued species lack the resilience of wild forests when pests, fire and other perturbations occur.

Trees have survived until today only because there is a great deal of genetic diversity within each species.
—PETER WOHLLEBEN, *The Hidden Life of Trees*

Ecosystem Stability in Diversity

An ecosystem is a complex community of producers, consumers, decomposers and detritivores, which interact within boundaries imposed by their physical surroundings to cycle energy and material through the web of life. In any ecosystem, the eaters and the eaten are joined through a web of interdependence. A kind of biological warfare is constantly waged between predator and prey, host and parasite, as each species jockeys for an upper hand. Mutations or new gene combinations conferring an advantage for one species are soon matched by a countering response in the other species to restore a balance. For example, a fungal parasite may develop an enzyme that digests the cell wall of a plant more efficiently, enabling it to penetrate its target species more readily. But in the population of its host organisms, individuals with thicker or tougher cell walls will be more likely to survive and reproduce. Over time, the parasite will have to come up with another innovation to penetrate the host's improved defences. So although there is a constant state of flux and change, the long-term overall effect is a standoff between the various constituents of ecosystems.

Tropical rainforests, believed by biologists to be home to most species on Earth, are a vast patchwork mosaic of diversity in which particular species are often severely confined by their habitat requirements to small areas within the forest. Agro-forestry expert Francis Hallé says that introduced species do not spread in tropical forests the way the purple loosestrife plant, for example, has exploded in North America, because the area of potential habitat is smaller and there are always many potential predators to keep any introduced species under control.

Just as genetic diversity confers resilience on a species, diversity of species within any ecosystem is also a factor in maintaining balance and equilibrium within that community of creatures. Species diversity, like genetic polymorphism within a species, appears to be an evolutionary survival strategy within whole ecosystems.

Across the broad expanse of the planet there exists a vast assortment of climatic and geophysical conditions—from the searing heat of deserts to the frigid cold of the permafrost above the Arctic Circle, from steamy equatorial river systems to dry grasslands, from the depths of the oceans to the soaring heights of rarefied mountains kilometres above sea level and to the inter-tidal junction between air, land and sea. Life has found ways to seize opportunities and flourish under all of these conditions.

"Extremophiles," organisms that live in Earth's most extreme places, show us the versatility of life. It seems there is no place on our planet devoid of life: NASA scientists revived a bacterium in 2005 that had sat dormant in a frozen Alaskan pond for 32,000 years; soil samples taken from the ocean's deepest point 11 kilometres down are filled with single-celled organisms called foraminifera; and entire communities of organisms, including clams, tube-worms and bacteria, thrive around hydrothermal vents synthesizing energy from chemicals in the water rather than from sunlight. Every ecosystem contains a variety of species—and each species possesses its locally confined set of genes. So even where species diversity is relatively limited—as, for example, in boreal forests—the genetic variation within a species in one watershed will differ from that within the same species in the next watershed. Every ecosystem is unique and special. Every ecosystem is local.

In this way, Earth itself is a mosaic of diversity within diversity, a patchwork of ecosystems, species and genes. Over time, this fabric of interconnections has been torn by major upheavals, most recently in North America by the rapid extermination of billions of passenger pigeons, millions of bison and vast tracts of shortgrass prairie and old-growth forests. The persistence of plants and animals after such catastrophic change is testimony to the tenacity of the planet's biodiversity.

Human Cultural Diversity

Human beings have extended diversity to yet another level. The successful evolution of our species has depended on the brain's gifts of memory, foresight, curiosity and inventiveness—and its recognition of patterns and cycles in the world around us. Our ability to exploit our surroundings and to pass on with language the lessons acquired by failure and success accelerated the pace of human evolution. Humans have had an added edge in culture. Every individual human being must begin life from the same

starting point, as an infant, laboriously acquiring all the accumulated lore and beliefs of society until he or she is ready to become
a productive adult. But culture grows steadily, without having to
go through the same learning curve every generation. Compared
with rates of biological change, culture evolves with lightning
speed—and for this reason we have come a long way in a relatively short time.

Using molecular techniques to measure degrees of biological
relatedness in DNA, scientists can identify the origins of human
beings and trace their movement across the continents. Population biologists have concluded that a mere 195,000 years ago,
the ancestors of all of humanity arose along the great Rift Valley of Africa. From there, they radiated out—northeast across
the Sahara, southwest into what is now South Africa, northward
across the Arabian Peninsula and west to India (Figure 6.3).

||||||||||||||||||||||||||||||||||||||||||||||||||||||||||||||||||||||||||||||||||||||||||||||||||||||

Figure 6.3: The spread of human beings across the planet.
The numbers indicate number of years ago.

Adapted from John Pickrell, "Instant Expert: Human Evolution," *New Scientist*, http://
www.newscientist.com/article.ns?id=dn9990. Accessed April 3, 2007.

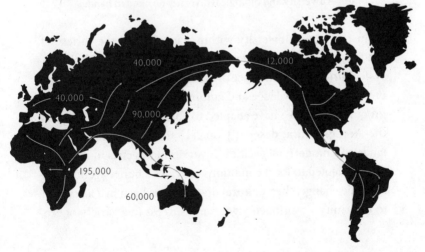

From these new locations they fanned out into Europe and Russia, from New Guinea to Australia, into Siberia and across the Bering land bridge to the Americas. Although people are wonderfully diverse in skin colour and facial and other physical features, the most significant differences between groups of human beings are not biological but cultural and linguistic.

In many animals, genetically encoded instinctive behaviour has enabled them to persist and survive. In contrast, the great strategy in our species has been the evolution of a massive brain capable of assessing sensory information and therefore deliberately making choices. Most of our instinctive behaviour has been replaced by flexibility, an ability to change patterns of behaviour on the basis of observation and experience. Culture and language have been our crucial attributes, enabling us to adapt to a wide range of surroundings and conditions. As Vandana Shiva has said:

> Diversity is the characteristic of nature and the basis of ecological stability. Diverse ecosystems give rise to diverse life forms, and to diverse cultures. The co-evolution of culture, life forms, and habitats has conserved the biological diversity of this planet. Cultural diversity and biological diversity go hand in hand.

Just as genetic diversity within a species and the variety of species within an ecosystem allow single species or whole ecosystems to survive in the face of changing conditions, so diversity of traditional knowledge and culture has been the main reason for our success. We have adapted to environments as diverse as the Arctic tundra, deserts, tropical rainforests, prairie grasslands and modern megacities. If variation of genes in a species that is adapted to local conditions provides a buffer against catastrophic change, then cultural diversity has been just as crucial to humanity's continued vigour and success in a variety of ecosystems. Ethnobotanist Wade Davis has defined the sum of all cultures, which have been so critical to human survival in so many different ecosystems, as the *ethnosphere*. Scientists are rightly concerned about the rapid extinction of species within the biosphere, but Davis points out that the threat that 50 per cent of all human languages may disappear by the middle of this century ought to arouse just as much concern.

One might suggest that the long and gnarled path of evolution might arrive at a point where the very "best" genetic or species combination or "ideal" human society has been achieved and should then spread globally to replace all "less advanced" forms. Diversity would then be totally outmoded. If global conditions were unchanging and uniform, it is at least theoretically possible that there might be a most highly evolved and stable society or species. But in nature, "best," "superior" and "advanced" are nonsensical terms because on Earth conditions are *never* constant. The nature of the biosphere—that thin layer of air, land and water within which life can be found—is that change, albeit often at a geological snail's pace, has always occurred, so there can never be one perfect or ideal state. Nature is in constant flux, and diversity is the key to survival. If change is inevitable but unpredictable, then the best tactic for survival is to act in ways that retain the most diversity; then, when circumstances do change, there will be a chance that a set of genes, a species or a society will be able to continue under the new conditions. Diversity confers resilience, adaptability and the capacity for regeneration.

The Living Planet

From genes to organisms to ecosystems to cultures—at every level the patchwork diversity adds up to a single living whole. The final sum may be Earth itself. Many cultures have myths in which the planet we inhabit is perceived as alive—as a creative force, a nurturing goddess or a collection of powerful spirits. And modern science may be providing corroborating evidence for such a view of life on Earth. When the first images of our planetary home were taken by astronauts in space, the beauty of the blue orb cloaked in white lace was breathtaking and changed our perception of Earth. This is our home, free of human borders and boundaries, a single integrated whole with a thin ephemeral layer within which life flourishes.

A scientific expedition from another galaxy in search of life in the universe might reasonably conclude from observing this planet that it is a living entity. The tenacious layer of protoplasm that wraps the Earth has survived and flourished through endless planetary upheavals. Continents have drifted around the globe, mountains have thrusted skyward, gaseous mixtures in the atmosphere have waxed and waned, and the temperature has fluctuated from tropical heat to the frozen grip of ice sheets. No life form managed to survive this turmoil on its own but depended on help from other organisms.

A single cell can be a complete organism, possessing all of the genetic material and molecular architecture to respond to the environment, grow and reproduce. Multicellular organisms such as sponges and slime moulds may have complex life cycles, yet when their individual cells are isolated each can grow and multiply as if it were a complete organism—or the cells can reassemble themselves into the multicellular aggregate that behaves as a single organism.

In fact, each cell in our bodies is an aggregate of species functioning as a single entity. In the 1970s biologist Lynn Margulis resurrected a theory that structures called organelles found within cells of complex organisms are actually the evolutionary remnants of bacterial parasites. Armed with the tools of molecular biology, she showed that organelles are able to reproduce within a cell and even possess DNA and distinct hereditary traits. So,

Margulis proposed, organelles were once free-living organisms that invaded cells and were eventually integrated into the host as mitochondria and chloroplasts. Giving up their independence, these microbial relics received nourishment and protection from the host cell. Thus, each of us is a community of organisms. We are each an aggregate of trillions of cells, every one of which is inhabited by numerous descendants of parasites; they now provide services for us in return for an ecological niche.

By sheer numbers, chloroplasts and mitochondria, rather than humans, are Earth's dominant life forms. Wherever we go, the mitochondria go too, since they are inside us, powering our metabolism: that of our muscles, our digestion, and our thoughtful brains.

—LYNN MARGULIS, *Symbiotic Planet*

Almost all of the 60 trillion cells that make up our bodies carry the entire genetic blueprint that specifies the development of a complete person. In principle, then, each cell has the potential, if triggered to read from the beginning of the instructions, to form another person or clone. Every cell may function according to the demands of the tissue or organ of which it is a part, just as every person may work according to the demands of his or her occupation. But each of us, like every cell, carries out many activities that we people do regardless of the job we have. As individuals, we cannot escape being part of families, communities or nations, which have their own characteristics and behaviour. Many other species are also part of larger groups.

Superorganisms

I once asked Harvard University's eminent biologist Edward O. Wilson, who died in 2021, why ants are so successful. He spent his entire career studying these ubiquitous insects, and he gave an animated response. Although the number of species of social insects is in the tens of thousands, there are millions of other non-social insects. But the social insects dominate the world because they behave, said Wilson, as a "superorganism."

> A colony of ants is more than just an aggregate of insects that are living together. One ant is no ant. Two ants and you begin to get something entirely new. Put a million together with the workers divided into different castes, each doing a different function—cutting the leaves, looking after the queen, taking care of the young, digging the nest out and so on—and you've got an organism, weighing about 10 kilograms, about the size of a dog and dominating an area the size of a house.
>
> The nest involves moving about 40,000 pounds of soil and sends out great columns of workers like the pseudopods of an amoeba, reaching out and gathering leaves and so on. This is a very potent entity. It can protect itself against predators. It can control the environment, the climate of the nest. When I encounter one of these big nests of leafcutter ants, I step back and let my eyes go slightly out of focus. And what you see then is this giant, amoeboid creature in front of you.

It was a thrilling description that lets us contemplate ants in a very different way.

In 1992, scientists in Michigan made the astounding announcement that the network of mycelia, threadlike extensions of fungi found in the ground, could be derived from a single individual, not an aggregate of different organisms. They reported a single organism that extended throughout 16 hectares! Amazingly, even that impressive record has been surpassed, and in a big way. In 2003, a root-rot fungus, *Armillaria*, covering 890 hectares was found in an Oregon forest.

When a person is part of a system, he cannot easily see what his role accomplishes... Unless he understands the system thoroughly, he will not have any inkling of the network of controls that may or may not exist to keep the flow(s) continuous, adapted to inputs, adapted to outside demands, and stabilized in the face of fluctuations.

—HOWARD T. ODUM, *Environment, Power and Society*

A grove of quaking aspen, the lovely white-barked trees whose leaves shimmer at the slightest puff of air, is, in fact, a single organism. Like a strawberry plant that sends out runners that put down roots and sprout leaves, quaking aspen multiply vegetatively. Shoots may grow up from a root 30 metres away. Thus, the aspen is another kind of superorganism that can exploit a diverse landscape—some parts may grow in moist soil and, through their common underground roots, share the water with other portions, perhaps growing in mineral-rich soil higher up. In Utah, a single aspen plant made up of 47,000 tree trunks was discovered. It covers an area of 43 hectares and is estimated to weigh almost 6 million kilograms.

So if at each level of complexity—cell, organism and ecosystem—new kinds of structures and functions emerge, then the total of all life on the planet can be taken as a single entity too.

A single envelope of atmosphere encircles the Earth, while water flows around the continents, creating great islands (Figure 6.4). The entire conglomerate of living things makes a wonderfully complex, interconnected community held together by the matrix of air and water. The entire layer of protoplasm (the living material within cells) on the globe is intermeshed into a living, breathing entity, which has survived through an immensity of time and space.

People are fond of applying mechanical metaphors to living systems: the heart is a pump, lungs are bellows, and the brain

|||||||||||||||||||||||||||||||||||||||||||||||||||||||||||||||||||||||||||||||||||||||||||||||||||||||

Figure 6.4: The continents as an island in a planet of water.

Adapted from a satellite portrait entitled "Our Spaceship Earth" (Burlington, ON: WorldSat International Inc., 1995).

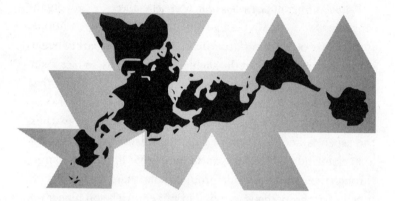

is a switchboard or computer; Earth itself is often referred to as a spaceship. But it is a mistake to compare living systems to machines. Mechanical devices constantly wear out with time unless they are carefully maintained and repaired by people. Living things persist on their own, healing, replacing, adapting and reproducing in order to continue. If the total of all life on Earth is a superorganism, then it must have processes that perpetuate its survival.

James Lovelock has called that totality of the living Earth Gaia, the ancient Greeks' name for Mother Earth.

Gaia is as indifferent to our fate as the stars. In the long run, the biosphere survives but its species do not... Virtually all of the species that have ever lived on this planet are now extinct. At times... half of the species on the planet have gone extinct almost at once. The next one hundred years may be such a time again. The story of life is punctuated by Ice Ages, volcanic winters, meteoritic collisions, mass dyings. And at the moment it is punctuated by us.

—JONATHAN WEINER, *The Next One Hundred Years*

Lovelock has pointed out that human activity is a major perturbation in the biophysical makeup of the planet. Many organisms can undoubtedly take advantage of the new conditions created by our disturbances. Life is opportunistic, and when a change occurs, life forms will be there to find a use for it. Thus, large tracts of clearcut forests often quickly "green up" with vegetation, and ungulates such as deer use the abundant foodstuff to grow and multiply. No doubt microbial species will flourish on our waste, just as gulls have a heyday in garbage dumps. But Gaia's feedback mechanisms take place over time, without regard to which species ultimately survive or disappear. The idea of Gaia, or the totality of the living Earth, may provide the comforting thought that life will survive the current spasm of human-induced extinction, but we should also remember that it will not ensure our own survival.

New Relationships

The intriguing hints and tantalizing clues emerging from the laboratories and the minds of modern scientists are creating a new story to give meaning and significance to our presence. We are creatures of Earth, created out of stardust, energized by the sun, carrying with us fragments of the first life forms—evidence of our kinship with every other creature on the planet. As Earth beings we share in life's basic survival method—diversity, both biological and cultural—and we are honed by evolution to live in the company of our fellow life forms. Armed with our emerging worldview, we find ourselves back on centre stage, holding the fate of our newfound family—and our own—in our trembling and incompetent hands.

In the cities inhabited by an increasing proportion of humanity, the links between human life and the lives of other creatures are often obscured by technology. Besides providing us with clean air, soil and water, other living organisms make our lives possible every day in countless fundamental ways. Every bit of our food was once living, however its source is disguised. Sugar, flour, vegetables, fruit, meat and spices nourish and delight us. When we clothe ourselves in cotton and wool, consume wood, plastics and fossil fuels, or fertilize our fields with manure, we are beneficiaries of once-living organisms. Insects fertilize plants that we depend on, horses and oxen provide muscle power, plants and animals are the source of many medicines. Body and soul, we are nourished by nature.

James Lovelock and the Concept of Gaia

James Lovelock began his career in medical research. In his quest to find ways to detect molecules in minute quantities, he developed an instrument so sensitive it could detect parts per *trillion*. Using the machine, he discovered CFCs in the atmosphere above Antarctica, thereby leading to the discovery that the ozone layer was being depleted.

In the early 1960s he was asked for advice in the design of the Surveyor spacecraft that was to explore the moon. Soon after, NASA asked Lovelock to design experiments for the Viking spacecraft that would search for life on Mars. Ruminating on the problem, Lovelock had to think about life itself, what it is and what distinguishes it from non-life. He realized that Mars and Venus have atmospheres composed almost totally of carbon dioxide, with no free oxygen. In contrast, Earth's atmosphere has small amounts of carbon dioxide and is 21 per cent oxygen. Although oxygen is a highly reactive element and tends to be removed from the atmosphere, plants continually release more oxygen to compensate for this loss.

What is remarkable is that the level of oxygen has remained relatively constant over a long period of time. A small increase to perhaps 25 or 30 per cent oxygen could cause the atmosphere to burst into flames, while a decrease to 10 per cent would probably be lethal to most life forms. Something has kept the amount of oxygen at just the right concentration for millions of years.

Lovelock reasoned that the oceans became salty by the leaching of minute quantities of salt from rock and soil into rivers and streams that flow to the sea. Why, then, haven't the oceans become saltier and saltier? Similarly, why haven't rising levels of carbon dioxide increased the temperature on Earth? On Venus, the carbon dioxide–rich atmosphere has turned the planet into an oven. In contrast, the thin atmosphere of Mars, which is low in carbon dioxide, cannot retain heat, and so the planet is frigid. Yet here on Earth the oceans haven't boiled away, even though the sun's intensity has increased by 25 per cent since the sun was formed. Something has kept the temperature of Earth and the salt concentration in the oceans relatively constant.

Lovelock's daring conclusion was that the total of all living things on Earth has somehow kept the concentration of carbon dioxide and oxygen, the amount of salt in the ocean and the surface temperature constant—not consciously or deliberately, but as part of an automatic process, just as our bodies increase our heart rate when we exercise or repair wounds when we are hurt. But now, technology has allowed us to generate massive quantities of greenhouse gases far faster than Gaia's capacity to remove them. Eventually, compensatory changes may reduce carbon dioxide levels, but not before tremendous ecological changes occur. Gaia's persistence plays no favourites on which species survive or disappear. □

Whereas our species once lived lightly on Earth, today we have exploded in numbers, technological dexterity and demand for consumer goods to such a degree that we are now co-opting much of the planet's productivity for our use. In the process, we deprive other species of habitat and opportunity and so drive them to extinction. Stanford University ecologist Paul Ehrlich's group estimated that human beings, one species among millions, now harness for our use 40 per cent of the net primary productivity of the planet. That is, of all the sunlight captured by plants, human beings deny a large portion of it to other species by using it for pasture, farmland, logging and so on. Our appropriation of this energy makes it unavailable for other species and drives them out of existence.

As we drain wetlands, dam river systems, pollute air, water and soil, clearcut vast tracts of forest, and develop land for agriculture, urban sprawl or industrial parks, the biodiversity that is the source of the planet's productive capacity is diminished. As a result, the world is experiencing a catastrophic rate of species extinction.

An indication of the unprecedented rate and scale of human activity is graphically illustrated by Alan Thein Durning in his paper "Saving the Forests: What Will It Take?":

Imagine a time-lapse film of the Earth taken from space. Play back the last 10,000 years sped up so that a millennium passes by every minute. For more than seven of the ten minutes, the screen displays what looks like a still photograph: the blue planet Earth, its lands swathed in a mantle of trees. Forests cover 34 percent of the land. Aside from the occasional flash of a wildfire, none of the natural changes in the forest coat are perceptible. The Agricultural Revolution that transforms human existence in the film's first minute is invisible.

After seven and a half minutes, the lands around Athens and the tiny islands of the Aegean Sea lose their forest. This is the flowering of classical Greece. Little else changes. At nine minutes—1,000 years ago—the mantle grows threadbare in scattered parts of Europe, Central America, China and India. Then 12 seconds from the end, two centuries ago, the thinning spreads, leaving parts of Europe and China bare. Six seconds from the end, one century ago, eastern North America is deforested. This is the Industrial Revolution. Little else appears to have changed. Forests cover 32 percent of the land.

In the last three seconds—after 1950—the change accelerates explosively. Vast tracts of forest vanish from Japan, the Philippines, and the mainland of Southeast Asia, from most of Central America and the horn of Africa, from western North America and eastern South America, from the Indian subcontinent and sub-Saharan Africa. Fires rage in the Amazon basin where they never did before, set by ranchers and peasants. Central Europe's forests die, poisoned by the air and rain. Southeast Asia resembles a dog with mange. Malaysian Borneo appears shaved. In the final fractions of a second, the clearing spreads to Siberia and the Canadian north. Forests disappear so suddenly from so many places that it looks like a plague of locusts has descended on the planet.

The film freezes on the last frame. Trees cover 26 percent of the land. Three-fourths of the original forest area still bears some tree cover. But just 12 percent of the Earth's surface—one-third of the initial total—consists of intact forest ecosystems. The rest holds biologically impoverished stands of commercial timber and fragmented regrowth. This is the present: a globe profoundly altered by the workings—or failings—of the human economy.

Seen this way, the planet's forests are being irrevocably lost in what amounts to a mere tick of the geological clock. Plotted over a mere ten millennia, the curve of forest devastation leaps almost straight off the page in our lifetime. And if we add to that graph the generation of pollution, loss of topsoil, increase in human numbers, production of greenhouse gases and so on, the curves all climb vertically in the very last moments. Individual disasters such as Chernobyl, large clearcuts, the explosion at Bhopal, the construction of megadams or oil spills are merely part of a terrifying spasm of annihilation.

... it is not Christ who is crucified now; it is the tree itself, and on the bitter gallows of human greed and stupidity. Only suicidal morons, in a world already choking to death, would destroy the best natural air-conditioner creation affords...

—JOHN FOWLES, quoted in T.C. McLuhan, *Touch the Earth*

Extinction Crisis

Our tenuous inferences about life in the past are based on fossil remains suggesting that species expand in number and complexity and then are suddenly reduced through successive spasms of extinction. Scientists have identified five major extinction crises over the past 500 million years, in which at least 65 per cent of all species known in the fossil record of the time disappeared. The fossil record is highly skewed—95 per cent of the quarter of a million known fossilized animal species are marine creatures. Nevertheless, these five major extinction episodes show groups of species disappearing on a massive scale, suggesting that the events were global. The Big Five were at the end of the Ordovician (440 million years ago), late Devonian (365 MYA), end Permian (245 MYA), end Triassic (210 MYA) and end Cretaceous (65 MYA). People often think that the dinosaurs were evolutionary losers because they suddenly disappeared, but the fact is that they ruled the land for some 175 million years. In contrast, our species has been around for less than 1 million years.

Following each major extinction, it has taken millions of years for the species that remain to branch out, expand in number and complexity, and restore the level of biodiversity that existed before each crash. In the words of Edward O. Wilson:

> The five previous major spasms of the past 550 million years... each required about 10 million years of natural evolution to restore. What humanity is doing now in a single lifetime will impoverish our descendants for all time to come.

We are fortunate to have evolved when biological diversity has been at the greatest level ever achieved. Succeeding human generations will not be as fortunate: the current extinction crisis is without precedent—never before has a single species been responsible for such a massive loss of diversity. In essence, humans are the catalyst driving Earth's sixth major extinction event.

When the first European settlers arrived in what is now the United States, the continent was covered by an estimated 3.2 million km² of forest. In just 500 years, all but 220,000 km² have been cleared.
—EDWARD GOLDSMITH et al., *Imperiled Planet*

By comparing the estimated rate of species loss today with the changes observed in the fossil record, Wilson concluded that the present extinction rate is "1000 to 10,000 times higher than existed in prehistoric times." Based on the current rate of destruction of tropical rainforests (about 1.8 per cent per year), about 0.5 per cent of all species are gone or going annually. More than half of all species live in tropical rainforests, and if we conservatively estimate that there are 10 million species, then the rate of extinction is more than 50,000 species a year—that's 137 a day, 6 an hour! This is an extremely conservative estimate, since it doesn't include species lost through pollution, non-clearcutting forest disturbances and the introduction of exotic species. Research published in 2019 by the UN Intergovernmental Science-Policy Platform on Biodiversity and Ecosystem Services found "1 million species already face extinction, many within decades, unless action is taken to reduce the intensity of drivers of bio-diversity loss. Without such action, there will be a further acceleration in the global rate of species extinction."

It is widely agreed that changes to biodiversity due to human actions have occurred more rapidly over the past sixty-five years than at any time in human history. According to the WWF *Living Planet Report 2020*, more than 1 million species face extinction—and population sizes for monitored mammals, birds, amphibians, reptiles and fish declined 68 per cent between 1970 and 2016. A 2019 UN report found more than 40 per cent of amphibian species, almost 33 per cent of reef-forming corals and more than a third of all marine mammals are threatened with extinction. A comprehensive U.S. study in 2021 found that many insect species worldwide are declining because of pesticides, light pollution, habitat destruction, climate disruption and more. In the North Atlantic, the biomass of larger fish at the top of the marine food chains (for example, cod) declined by two-thirds during the second half of the twentieth century alone and by a factor of nine during the entire century. In 2003, a paper in the journal *Nature* revealed the urgency: only 10 *per cent* of all the large fish—including open-ocean species such as tuna and marlin, and groundfish such as cod and halibut—remain in our oceans. Most alarming is our lack of restraint when new fish communities are discovered. The same study, which took ten years to compile, also showed that it took industrial fishers only ten to fifteen years to deplete fish communities to one-tenth of their original size. It is no wonder that the late Nobel laureate Paul Crutzen dubbed our epoch the Anthropocene, after the humans that have had a significant impact on the Earth's ecosystems and climate. The most frightening aspect of the current extinction crisis is our ignorance of and lack of concern for what we are losing. According to John A. Livingston:

> We have seen the bison, the trumpeter swan, and the bighorn sheep fall before the gunners; we have seen the prairie dog, the black-footed ferret and the whooping crane give way before the sod-busters; we have seen the giant baleen whales reduced to the vanishing point by international commercial greed. Most significant of all, perhaps, has been the unchanging traditional assumption that although the loss of these animals may well have been regrettable, it was inevitable and unavoidable in the context of the advancement of human progress...
>
> What is relevant is that, if for no other reason than his own survival, man must soon adopt an ethic toward the environment. "The environment" encompasses all nonhuman elements in the one and only home we have on Earth.

Preserving the Web of Life

Although extinction is as necessary to the evolutionary process as species formation, it has accelerated at an unprecedented rate as a result of human depredation. There are many reasons to be alarmed by the loss of species, all of them completely selfish. Perhaps the shallowest is regret for loss of species whose potential utility for humankind is yet to be discovered. Another is that species such as spotted owls or marbled murrelets serve as "indicator species" of the state of the planet, just as canaries did for the state of air in coal mines. In other words, when such species disappear, they indicate that the planet as a whole may have become less habitable in a way that may be relevant to humanity.

The emerging viruses are surfacing from ecologically damaged parts of the Earth. Many of them come from the tattered edges of tropical rain forest or tropical savanna that is being settled rapidly by people. The tropical rain forests are the deep reservoirs of life on the planet... [including] viruses, since all living things carry viruses. In a sense, the Earth is mounting an immune response against the... flooding infection of people, the dead spots of concrete all over the planet...
—RICHARD PRESTON, *The Hot Zone*

As a biologist, I find it much more compelling to regard the current makeup of life on Earth as the latest stage in evolution—the reason the planet is as productive as it is. Even though we have little understanding of what the components of this complex web of life are, we know with absolute certainty that it is the web as a whole that has made it possible for human beings to exist. To tear at the web in such a massive way with so little regard for our own future is a kind of collective insanity that is suicidal.

The COVID-19 pandemic illustrates the consequences of tearing at that web. Most "novel pathogens" to which we haven't developed immunity are zoonotic, including Ebola, Zika, West Nile virus, SARS, HIV and others. That is, they jump from other animals to humans. In 2021, a coalition of health and conservation organizations pointed to evidence showing that disease outbreaks like COVID-19 are caused largely by "increasing rates of deforestation and land-use change due to population growth and urbanization—coupled with growing globalization and excess production driven by consumerism."

Our major crises—pandemics, climate disruption and biodiversity loss—all have roots in our lack of recognition of our place in nature.

In 1990, the Worldwatch Institute designated the next ten years the Turnaround Decade, the period during which it was essential to shift the trajectory of human activity to a sustainable level. The 1990s passed by, and now, well into the new millennium, the planet is exhibiting increasingly troubling signs of stress. Many of us are alarmed and have been trying to find the best strategy for action. The late famed American environmentalist David Brower called for a program of CPR for the planet. Brower's CPR stands for "Conservation, Protection and Restoration" and is deliberately used as a reminder of cardiopulmonary resuscitation, which was the original source of the acronym. Brower once told me that he believed restoration must be our priority in the years to come, and I agree.

But how? Science provides tiny, fragmented insights into the natural world. We know next to nothing about the biological makeup of Earth's life forms, let alone how they are interconnected and interdependent. Nor do we understand the physical features and complexity of the atmosphere, land masses and oceans. It is a dangerous delusion if we think we know enough to "manage" forests, climate, water or wild ocean or land animals.

Since after extinction no one will be present to take responsibility, we have to take full responsibility now.
—JONATHAN SCHELL, *The Abolition*

Extinction, of course, is irreversible. And even heroic measures to keep an endangered species going don't stand much of a chance without profound changes in human behaviour and genuine protection of the species' habitat.

The thin layer of biological complexity within the biosphere ensures the productivity and cleanliness of the soil, air and water. Only time and nature safeguard these life-supporting elements and keep them intact. Remarkably, if we pull back and decrease or halt our assault on a given environment, nature can restore itself. We have seen life return to Lake Erie, once declared "dead" from eutrophication, vegetation revive around Sudbury after sophisticated scrubbers were installed to reduce acidic emissions from smelter stacks, and fish reappearing in the River Thames in England after antipollution laws were imposed.

A Day in the Life

Economic growth is considered necessary to satisfy the needs of all members of society. But this growth is at the expense of the rest of life on Earth, and it behooves us to reflect on what best satisfies our needs and brings us happiness. I was able to do that in 1989, when my six- and nine-year-old daughters, my wife and I were guests of the Kayapo leader Paiakan in the village of Aucre, deep in the Amazon rainforest. For ten days, we lived a simple life, sleeping in hammocks stretched inside a mud hut. The nearest settlement was a fourteen-day canoe trip, and the two hundred residents of Aucre had no plumbing, tap water or electricity. The pace of life was leisurely. Often we awoke to find a roomful of children inches away, observing us. We were obviously the entertainment for that non-television-watching audience. Breakfast might be bananas or guavas and leftovers from the night before. We would drink fresh water from a spring and meet socially with others for a long morning swim while the children and women fished for a delicious fish they called *piaau*.

Each day we went on expeditions through the forest to gather fruits and edible plants or travelled by dugout canoe in search of fish, turtle eggs or capybara. In the village, we witnessed a spectacular three-day festival to celebrate women and their

Those who contemplate the beauty of the Earth find reserves of strength that will endure as long as life lasts. There is symbolic as well as actual beauty in the migration of birds, the ebb and flow of tides, the folded bud ready for spring. There is something infinitely healing in the repeated refrains of nature—the assurance that dawn comes after the night and spring after the winter.

—RACHEL CARSON, *Silent Spring*

fertility, observed an emotional funeral for an old man who had died of tuberculosis and watched the men weave straps to carry babies, or feather headdresses. There was time to reflect, play, observe and learn. My daughters wept when our ten-day visit was over and we had to leave.

What a contrast with our daily life in the rich, industrialized country of Canada. Our time is set by obligations and commitments—the clock and daily schedules dictate every activity. Days are fragmented into short intervals that preclude any time for observation or reflection. As a boy, I loved to read articles about the world of the future when robots and machines would serve our every need and free us to read, play and interact with others. Well, that future arrived some time ago. In my home, I now have computers, smartphones, a dishwasher, and a clothes washer and dryer. But life has accelerated as we race through it, and there is little time to watch and think. Thinking back to our time in Aucre, I often ask myself what this way of life and all of the material things are for. Am I happier or freer now than when we were swimming in the river, fishing or singing in Aucre? Before my children got caught up in the turbulence of the adult world and economics, they knew the answer to this question. That's why they wept when we left Aucre. □

Even though we can't re-create what no longer exists, there are things we can do to stimulate the natural process of regeneration. First we must rein in our destructive ways and then provide conditions to encourage the return and regrowth of life. We can liberate land and creeks from rubbish, concrete or asphalt, cultivate specific vegetation and even reintroduce plant or animal species that were once present. But mainly, we must give Earth's restorative powers time to act. There are projects that could be inspirational models for beginning to heal the planet. From Japan to Canada, people are working to "daylight" creeks and rivers— that is, to re-expose water systems that have been buried under urban development. Once the water is opened to the air, freed from concrete coffins, allowed to flow across soil and surrounded by plant life, it can support life again and purify itself and its surroundings. Australians have also undertaken an economic and ecological analysis indicating that it would be practical to take down a bitterly opposed dam that flooded the Pedder River in Tasmania forty-five years ago. In the United States, wolves have been reintroduced into Yellowstone National Park, while free-ranging herds of bison are being returned to parts of Montana and Wyoming.

In small ways as well as large, there are signs that we are turning away from destroying natural systems. Native flora are replacing exotic, chemical-dependent, high-maintenance lawns and bedding plants on public and private land in cities and towns across Canada, providing habitat for insects, birds and small mammals. Organic farming is beginning to become an economic alternative, as demand for pesticide-free produce grows, allowing soil organisms to thrive and multiply in the service of productivity. And individuals are becoming involved in small local conservation projects such as butterfly flyways that enable many forms of life to coexist with human beings.

In the past, it was possible to destroy a village, a town, a region, even a country. Now it is the whole planet that has come under threat. This fact should compel everyone to face a basic moral consideration; from now on, it is only through a conscious choice and then deliberate policy that humanity will survive.

—POPE JOHN PAUL II, "The Ecological Crisis: A Common Responsibility"

Humanity has shown itself capable of heroic acts of courage and sacrifice in times of crisis. When the Japanese attacked Pearl Harbor on December 7, 1941, North Americans knew that life would never be the same. They didn't debate economic cost; they knew they had to do whatever it took to win—and they did. The ecological holocaust that has been loosed on the planet is the equivalent of "a million Pearl Harbors happening at once," in Paul Ehrlich's words. The challenge is to make the extinction threat as real as Pearl Harbor.

From his work studying ants of the world, Edward O. Wilson offered this humbling perspective:

If we were to vanish today, the land environment would return to the fertile balance that existed before the human population explosion. But if the ants were to disappear, tens of thousands of other plant and animal species would perish also, simplifying and weakening the land ecosystem almost everywhere.

In the end, the crucial change is attitudinal; we have to see ourselves in a different relationship with the rest of nature.

THE CANTICLE OF BROTHER SUN
Most high, omnipotent, good Lord
To you alone belong praise and glory
Honor, and blessing

No man is worthy to breathe your name.
Be praised, my Lord, for all your creatures.
In the first place for the blessed Brother Sun
Who gives us the day and enlightens us through you.
He is beautiful and radiant with his great splendour,
Giving witness of you, most Omnipotent One.
Be praised, my Lord, for Brother Wind
And the airy skies, so cloudy and serene;
For every weather, be praised, for it is life-giving.
Be praised, my Lord, for Sister Water
So necessary yet so humble, precious, and chaste.
Be praised, my Lord, for Brother Fire,
Who lights up the night,
He is beautiful and carefree, robust and fierce.
Be praised, my Lord, for our sister, Mother Earth,
Who nourishes and watches us
While bringing forth abundant fruits with coloured flowers
And herbs.
Praise and bless the Lord.
Render him thanks.
Serve him with great humility. Amen.

—SAINT FRANCIS OF ASSISI

7

The
Law of Love

Being a human being—in the sense of being born to the
human species—must be defined also in terms of becoming a human
being... a baby is only potentially a human being, and must grow
into humanness in the society and the culture, the family.
—ABRAHAM H. MASLOW, *Motivation and Personality*

AS BEINGS WHO emerged from and are formed by the
elements of the Earth, our very existence is absolutely
dependent on air and sunlight to kindle our metabolic furnaces, water to facilitate and give form to life's processes, and
soil to provide the atoms and molecules that enable cells to grow,
replace themselves and reproduce. These foundations of all life
are enriched and maintained by the totality of life's diverse forms.
Together these factors set the real bottom line, the needs that
must be met for us to live. Our bodies reflect the importance of
those needs with fine-tuned physiological alarms that impel us
to obtain air, water, soil and energy when they are needed. Our
ability to grow and flourish is directly related to the quantity and
quality of these fundamental requirements.

But human beings do not live by bread alone. The distinguished psychologist Abraham Maslow pointed out that fulfilling

our basic physiological requirements is our most urgent need and dominates our thinking and behaviour. When air, water, food and warmth are in adequate supply, however, they fade from our thoughts; then another constellation of needs emerges that is just as crucial to our well-being:

> What happens to man's desires when there is plenty of bread and when his belly is chronically filled? *At once other (and higher) needs emerge*... and when these in turn are satisfied, again new *(and still higher) needs emerge*... basic human needs are organized into a hierarchy of relative prepotency.

Notwithstanding that many of us who share this planet do not have bread in our bellies, when these "higher order" needs are met, we are able to reach our full potential; both our physical and our psychic health and well-being depend on this set of basic needs.

... as an animal [a person] must breathe, eat, excrete, sleep, maintain adequate health, and procreate. These basic needs constitute the minimum biological conditions which must be satisfied by any human group if its members are to survive. These physiological or biogenic needs and their functioning interrelations constitute the innate nature of man.
—ASHLEY MONTAGU, *The Direction of Human Development*

We are social beings—herd animals who depend on each other at every stage of our lives. Like many other animals, we are born unable to care for ourselves; we need a long period of care from our parents so that we can grow and learn in safety. As each of us develops, we need companions to define and extend our sense of self, and a community in which we find opportunities for a mate, for rewarding activity and for conviviality. These needs are absolute, inalienable, and where they are not met we suffer, even perish. Like the caribou that wanders too far from the herd, we cannot thrive in isolation from our kind. From the very beginning of life each one of us is shaped for and by close relationships with other human beings.

The Prime Directive

In every human society one overwhelming need directs the development of every individual. According to Ashley Montagu, for an adequate healthy development,

> the human infant requires, beyond all else, a great deal of tender loving care. Health at a very minimum is the ability to love, to work, to play, and to think soundly... The infant's need for love is critical, and its satisfaction necessary if the infant is to grow and develop as a healthy human being.

Numerous studies indicate that love is an essential part of a child's upbringing from birth; it helps the individual to thrive, while it teaches the qualities necessary for belonging to a wider community. Being loved teaches us how to love, how to imagine and feel for another person's existence, how to share and co-operate. Without these skills, how long could any group of humans survive together? In its purest form, the bond between parent and infant illustrates love's remarkable property of reciprocity. The joy of unconditional parental love is fully returned by the object of that love.

This mutual attraction may be built into the very structure of all matter in the universe. Love may in truth make the world go round—or at least hold it together.

As energy from the natal cauldron of the Big Bang 15 billion years ago filled the ever-expanding universe, the newly formed particles that would eventually coalesce into atoms felt a mutual attraction even as they rushed away from each other. A body with mass tugs at any other body with mass. When protons and electrons appeared, this attraction based on mass was greatly amplified by the pull of opposite electric charges. Throughout the universe, however imperceptibly, all matter feels drawn together.

Galaxies suddenly appeared a billion years after the Big Bang. Long after our own Milky Way galaxy and our sun had evolved, hydrogen had transformed itself into living matter, in the form of cells, on Earth. A membrane demarcates a cell from its surroundings, forming a barrier that allows materials to be concentrated within the cell and enables metabolism to take place. Even though a membrane separates life from its environment, membranes have such a strong affinity for each other that when two cells are brought close together they fuse, and the cytoplasmic contents of both cells are combined into one. Viruses, bacteria and protozoa, such as the amoeba and paramecium, fuse in the act of genetic recombination. All flora and fauna endowed with sexual cycles are drawn together in that wondrous act of biological reproduction.

In trees and plants one may trace the vestiges of amity and love... The vine embraces the elm, and other plants cling to the vine. So that things which have no powers of sense to perceive anything else, seem strongly to feel the advantages of union.

But plants, though they have not powers of perception, yet, as they have life, certainly approach very nearly to those things which are endowed with sentient faculties. What then is so completely insensible as stony substance? Yet even in this, there appears to be the desire of union. Thus the lodestone attracts iron to it, and holds it fast in its embrace, the attraction of cohesion, as a law of love, takes place throughout all inanimate nature.

—DESIDERIUS ERASMUS, quoted in P. Crean and P. Kome, eds., *Peace, a Dream Unfolding*

When we observe the care with which a mud dauber prepares a mud enclosure, inserts a paralyzed victim as food and deposits an egg, can we be so anthropocentric as to deny this the name of love? How else could we interpret the male sea horse's protective act of accepting babies into his pouch, the months-long incubation of an emperor penguin's egg on the feet of its vigilant parent or the epic journey of Pacific salmon returning to their natal stream to mate and die in the creation of the next generation? If these are innate actions dictated by genetically encoded instructions, all the more reason to conclude that love in its many manifestations is fashioned into the very blueprint of life.

In experiments that would be frowned on today, H.F. Harlow and M.K. Harlow performed classic experiments with baby monkeys that had been taken away from their mothers shortly after birth. When given a choice between a wire form in which food was offered and a soft, terry cloth–covered figure that had no food, the babies preferred the surrogate cloth form, to which they could cling, even though the wire figure was their source of food. Although well fed and cared for physically, the monkeys exhibited abnormal behaviour when they grew up, including a complete lack of interest in raising young. Thus, experiments with primates reveal both the powerful need for love—where the merest hint of a loving parent was chosen in preference to food— as well as the tragic, persistent consequences of deprivation. In a universe designed according to the principles of mutual attraction, co-operation and coherence—Erasmus's "law of love"—we humans, who are even more highly socialized than monkeys, have a fundamental need to love and be loved. As Ashley Montagu observes:

... the biological basis of love lies in the organism's ever-present need to feel secure. The basis of all social life has its roots in this integral of all the basic needs which is expressed as the need for security, and the only way in which this need can be satisfied is by love... The emotional need for love is as definite and compelling as the need for food... in order that he [man] may function satisfactorily on the social plane, the most fundamental of the basic social needs must be satisfied in an emotionally adequate manner for personal security and equilibrium.

Love shapes us even before birth. Secure in the equilibrium of the womb, a fetus is exquisitely attuned to the physiological, physical and psychological state of its mother. In turn, its growth and development within the womb affects the sequence of hormone-controlled changes in the mother's body during the pregnancy. Mother and child are entwined in a collaboration. According to Montagu:

> ... the fetus is capable of responding to sound as well as to pressure, and the beating of its own heart at about 140 beats a minute, together with the beating of its mother's heart at a frequency of 70, provides it with a syncopated world of sound. Laved by the amniotic fluid to the symphonic beat of two hearts, the fetus is already in tune with the deepest rhythms of existence. The dance of life has already begun.

From its beginning to its final faltering steps, that dance continues to be deeply interactive. After birth, breastfeeding continues the intimate connection between mother and baby. At the baby's cry, even at a distance, the mother's breasts "let down" their milk. And like the love it fosters and expresses, the benefits of nursing are reciprocal. The act of sucking not only provides nourishment and stimulates the baby's oral regions, but also activates the alimentary, endocrine, nervous, genitourinary and respiratory systems in the infant. At the same time, the infant's touch induces maternal contractions of the uterus, helping it to regain its normal size and shape and reducing afterpains and bleeding from the uterine lining. No doubt the skin contact and nursing together induce the release of endorphins to create a state of well-being and happiness in both mother and child. Alfred Adler believed that this state was responsible for the continuation of the species:

> The first act of a new-born child—drinking from the mother's breast—is co-operation, and is as pleasant for the mother as for the child... *We probably owe to the sense of maternal contact the largest part of human social feeling, and along with it the essential continuance of human civilization.*

Love is the defining gift that confers health and humanity on each new human; it is the gift that passes on endlessly, given and given again by each generation to the next. In the words of Ashley Montagu:

> By being loved the power is released in the infant to love others. This is a critically important lesson that, as human beings, we need to understand and learn: that the cultivation of the growth and development of love in the child should be its natural birthright.

Love Potions

When we first meet a new love, we often talk of there being a "chemistry" that draws us together. Research is now showing that this is more than just an expression—it is a scientific truth: our bodies are having their own love affair, despite what we think our hearts feel or our intellects need. Love naturally stimulates myriad biochemical pathways in our bodies, delivering chemicals that give us the feeling of love in all its manifestations, from the exhilaration of being "madly in love" to the contentment and comfortable familiarity of long-term relationships.

Scientists now think there are three stages of love, each defined by a different cocktail of brain chemicals that influence our emotions and actions. The first stage is best described as lust. In this phase, the sex hormones testosterone and estrogen do the talking, with a chaser of "feel-good" endorphins. This powerful drive for sex is evolution's way of motivating individuals to sustain a population, but hot sex with multiple partners does not necessarily a good society make. Rather, there ideally needs to be relationships between individuals that are sustained long enough to complete parenting duties, at least for a while. Thus, we enter stage two of love: the arena of romantic love.

In love's second phase, there is intense attraction, sometimes called obsessive love, between people. We can't sleep or eat, butterflies set up shop in our stomachs, our palms sweat and at times it feels as if we navigate our days in a haze of love-struck distraction. Here we begin to refine the feelings of lust. We enter emotional terrain where our body delivers a new mix of chemicals, meted out to create the heady bliss of love.

Anthropologist Helen Fisher and her colleagues gave brain scans to people who declared they were "madly in love." When shown pictures of the subject of their passion, the parts of the brain associated with pleasure—those rich with receptors for the neurotransmitter dopamine—lit up.

In the right proportions, dopamine creates intense energy, exhilaration, focused attention, and motivation to win rewards. It is why, when you are newly in love, you can stay up all night, watch the sun rise, run a race, ski fast down a slope ordinarily too steep for your skill. Love makes you bold, makes you bright, makes you run real risks…

—DR. HELEN FISHER, quoted in "Love: The Chemical Reaction"

Romantic love exhilarates us because brain chemicals kick in a feedback of motivation and reward. When dopamine levels are high, we focus in on the stimuli that make us feel good. We are highly motivated to gain our "reward"—in this case, the elated feelings we have when with our love. During the phase of romantic love, dopamine delivers extreme pleasure, while norepinephrine and serotonin provide the excitement—thus the racing heart and the sweaty palms. This hormonal triumvirate is controlled by phenylethylamine, or PEA, which facilitates our infatuation. In its chemical structure, PEA is similar to amphetamines; it is a natural upper that fuels this intense period of love. We crave these chemicals and the feelings they deliver; we feel "high" when in love. But, as with most drugs, our body can build up a tolerance and we want more and more. Herein lies the rub. We'd all be love junkies—forever seeking the elusive high (read: having a series of short relationships to fuel our unquenchable desire)—if it weren't for the fact that long-lasting relationships eventually morph into a third phase of love where there is yet another shift in brain chemistry. (Some relationships, of course, never make it to stage three, which happens about two to four years into the partnership. At this point, intense, passionate love can wane somewhat as the haze of infatuation lifts and you begin to see your partner in a new, more rational, light.)

So when the honeymoon is over, what then? In the third stage of love, dopamine is no longer in the driver's seat. Instead, the "cuddle chemical," oxytocin, takes the wheel. Oxytocin is a hormone associated with attachment and bonding. It enhances our connection to others and makes us feel calm, comfortable and secure. Both partners release oxytocin during sex, thus intensifying their attachment. Oxytocin also bonds parent to child. It is released when a mother nurses her baby and also when we hug those we love most. Oxytocin is nature's "glue" hormone that seals our emotional bonds and helps us stay together.

A study of prairie voles demonstrates the powerful role of oxytocin and another hormone, vasopressin, in social bonding.

Prairie voles are one of the 3 per cent of mammal species that are monogamous. (Humans are generally *not* considered part of this exclusive club.) Before mating, prairie voles interact freely with males and females. Eventually, a vole selects a partner, and during an intense period of mating, oxytocin and vasopressin are released. Forevermore, the male is indifferent to all other females and aggressively guards his mate. The couple groom each other for hours on end. When pups are born, both parents are affectionate and attentive. It is the rush of hormones that sealed their partnership for life. In further studies where vasopressin and oxytocin were blocked, the voles' interactions became fleeting, and no long-term attachment resulted.

The Family and Beyond

The fundamental unit that fosters and strengthens parental love is the family. It is an extraordinarily diverse human grouping, varying from the nuclear family of recent years in the West to the large extended families common in many parts of Africa to the collective kibbutzim of Israel. Polygamy, polyandry, the authority of the wife's brother, the power of the mother-in-law— whatever the shape of the family, one measure of its success is the happiness of its members. Sociologists have long known that there is no correlation between happiness and social class, per capita consumption or personal income. Instead happiness depends, apparently, on intimate human relationships. One consistent finding is that there is a much higher percentage of happy people among married couples than among those who have never been married, while divorced people are less happy than either of the other two groups. It seems that our "ever-present need to feel secure," which is satisfied by love, is as crucial to happiness among adults as it is among babies. Love is attraction, connection, coherence; it is a place to belong, a series of intersecting circles that extend out from each individual to include everyone else in varying degrees of closeness.

No man is an Island, *entire of itself; every man is a piece of the* Continent, *a part of the* main; *if a* Clod *be washed away by the* Sea, *Europe is* less, *as well as if a* Promontory *were, as well as if a* Manor *of thy* friends *or of* thine own *were; any man's death diminishes me, because I am involved in* Mankind; *And therefore never send to know for whom the bell tolls; it tolls for thee.*

—JOHN DONNE, *Devotions Upon Emergent Occasions*

Each of us is an expression of both our internal genetic makeup and the external experiences that create our life story. Each of us is part of a group, as well as being a unique individual: we are, you might say, both nature and nurture interacting, and it is often hard to disentangle the two. For example, I was born and raised in Canada, but my physical features reflect my pure Japanese genetic makeup. During the Second World War, my physical resemblance to the Japanese enemy made it difficult for people to remember that Japanese Canadians had not inherited an allegiance to the country of their genetic origins. The consequent upheaval in my life—evacuation from Vancouver, incarceration in remote camps in the Rocky Mountains and subsequent expulsion from British Columbia—shaped my personality and behaviour, illustrating the interaction of heredity and environment. Each of us has unique experiences that reflect our differences in gender, religion, ethnicity or socio-economic background, and the sum of those experiences moulds us into the kind of adult we become. The challenge is to create the kind of society in which our potential can blossom to the fullest extent. And in Montagu's opinion, that kind of society depends on raising healthy children:

> The child is the forerunner of humanity—forerunner in the
> sense that the child is the possessor of all those traits that, when
> healthily developed, lead to a healthy and fulfilled human being,
> and thus to a healthy and fulfilled humanity.

For Montagu, a healthy human being requires much more
than satisfaction of physiological needs in childhood. He lists the
following psychic needs of a growing young child that must be
fulfilled to ensure full development of a child's potential:

1. The need for love
2. Friendship
3. Sensitivity
4. The need to think soundly
5. The need to know
6. The need to learn
7. The need to work
8. The need to organize
9. Curiosity
10. The sense of wonder
11. Playfulness
12. Imagination
13. Creativity
14. Open-mindedness
15. Flexibility
16. Experimental-mindedness
17. Explorativeness
18. Resiliency
19. The sense of humour
20. Joyfulness
21. Laughter and tears
22. Optimism

23. Honesty and trust
24. Compassionate intelligence
25. Dance
26. Song

The extent to which families and their communities can fulfill those needs is the measure of their collective richness and vigour as a society. Beyond the essential bond between parent and child, human beings need to interact with others of their species. We are profoundly social animals, not atomized individuals moving freely and separately from all else. We derive our history, identity, purpose and ways of thinking from the social grouping in which we are born and raised and on which we depend.

Man! The most complex of creatures, and for this reason the most depen-dent of creatures. On everything that has formed you you may depend. Do not balk at this apparent slavery... a debtor to many, you pay for your advantages by the same number of dependencies. Understand that inde-pendence is a form of poverty; that many things claim you, that many also claim kinship with you.

—ANDRÉ GIDE, *The Journals of André Gide*

The Power of Touch

With the first kiss of a mother's lips, the first gentle whisper from a father, a tentative snuggle from a sibling, love works its way into our central core. Within a newborn's tiny brain, chemistry is afire, forging new neural pathways as the infant's senses of sound and smell and touch relay a flood of information from its new world.

The brain grows at an astonishing rate when we are young. We know, for instance, that young children can learn languages much more readily than adults. The rational brain is learning and absorbing, but so is the emotional brain. Developing infants need to bond with others, to feel security, compassion and love. The role of touch is paramount in human development. Without it, we lack the emotional nourishment we need to thrive, and, perhaps, even to survive.

Our body's largest organ—our skin—is primed for touch. Skin is riddled with nerves, touch receptors with which we feel hot and cold, pain and pleasure, tingles and tickles. We know from studies of primates and other animals that touch is central to proper emotional, psychological and physical development. In one study with monkeys, mothers were separated from their infants by a glass partition. In one group, the mother and child could still see, smell and hear one another, but they could not touch. The conditions were the same for a second group except that the pair could touch one another through holes cut in the partition. The babies who could not touch their mothers cried and paced, whereas the babies who could touch did not show serious behavioural problems. When reunited, the babies that had been deprived of touch clung obsessively to their mothers and failed to develop the level of independence and confidence shown by the other monkeys.

Not only may a parent's touch comfort and please its offspring, but new studies show that touch actually modifies brain development. Like human parents, rat parents have different parenting styles, with some being more attentive than others. Michael Meaney of McGill University demonstrated that differences in maternal care can make physical changes in the brain and determine how rats cope with stressful situations.

Meaney found the offspring of rats that spent more time licking and grooming their pups were better able to deal with stress later in life. The higher the incidence of grooming, the lower the levels of stress hormones produced by the pups. This meant that pups raised by particularly attentive mothers were calmer during stressful times later in life, and they also showed a greater capacity for learning. Overall health improved as well, since long-term exposure to high levels of stress hormones can contribute to chronic problems such as heart disease or diabetes.

The stimulation provided by licking actually triggered a change in the DNA chemistry in certain genes in the baby rats. As the mother licked and groomed her pups, she essentially "flipped a switch," turning on genes that reduced the amount of hormones released during periods of stress. More licking resulted in more receptors in the pups' brains that regulate the production of stress hormones.

Today, because we know about the critical role human touch plays in proper development, newborn babies often share a room with their mothers, we teach infant massage and parents are encouraged to carry babies close to the body in baby slings. Numerous studies clearly demonstrate that human development is enhanced by touch. Babies who are touched are alert, aware, active and engaged.

So great is our need for touch that touch therapy is becoming standard in neonatal care and is particularly helpful in the nurturing of premature babies. "Kangaroo Mother Care," for example, was developed in Colombia by pediatrician Edgar Rey. The idea of KMC is simple—preterm infants are held skin-to-skin on their mother's, or another caregiver's, bare chest. At first, KMC was developed in response to a shortage of incubators, but it is now clear that "human incubators" help children thrive. In the tiniest of infants, Kangaroo Care helps stabilize temperature, breathing and heart rates. These cuddled babies sleep longer, gain more weight, cry less often and are alert for longer periods. The constant contact calms the children, enhances the mother-infant bond and usually results in an earlier discharge.

Touch can bond child to parent, but, of course, it is a reciprocal exchange. We know that oxytocin surges when mothers give birth, nurse and care for their infants, but mammalian fathers can also experience hormonal fluctuations. Biologist Katherine Wynne-Edwards studied Djungarian hamsters, a species with particularly attentive fathers (unlike most mammals). These hamster dads are very involved in birth, from pulling babies from the birth canal to opening the baby's airways. As they do so, their estrogen and cortisol levels increase. If these hormones are suppressed, the males show decreased paternal involvement in the days following birth.

Wynne-Edwards also found that hormone changes can happen in other mammals, including humans. In one study, men about to become fathers for the first time were found to have lower levels of testosterone and cortisol and higher levels of estrogen (specifically estradiol) than men in the general population. Estradiol is a hormone that influences maternal behaviour.

The language of love is written without conscious thought in our bodies. Triggered by touch, but also by swirls of other sensory information that bathes us, our bodies respond automatically, binding us to others through shared experience. As we interact with our children, we are teaching what it means to be loved. When these experiences are inhibited, we are all the poorer for it.

Lessons Learned From Tragedy

Growing up without love has terrible repercussions for physical and social well-being. As Maslow notes:

> It is agreed by practically all therapists that when we trace a neurosis back to its beginnings we shall find with great frequency a deprivation of love in the early years. Several semi-experimental studies have confirmed this in infants and babies to such a point that radical deprivation of love is considered dangerous even to the life of the infant. That is to say, the deprivation of love leads to illness.

Unfortunately, humanity's capacity for love is counterbalanced by an awesome and awful capacity for brutality. When we see what happens to the victims of brutality, we realize the critical importance of love and its most likely source, the family. Scientific exploration of the effects of deprivation on young animals, such as the experiment described earlier, offends public sensibility—for good reason—today. But children in war-torn countries around the world have been subject to severe deprivation, and scientists have been able to study those children.

After the execution of dictator Nicolae Ceaușescu on December 25, 1989, we learned of the terrible plight of children institutionalized in Romania. Intent on increasing Romania's population, Ceaușescu had created a generation of unwanted children who were often abandoned to the state. An estimated 100,000 to 300,000 children were in institutions at the time of Ceaușescu's fall. Overcrowded and understaffed, most institutions provided little more than subsistence levels of food, clothing and shelter.

Among the seven hundred residential institutions for children, one type, the *leagane,* was for children who were not orphans but were abandoned or left for long periods by their parents. Scientists who examined them found rows of cots in huge dormitory rooms for the children, and staff members so harried they had no time to toilet train children or teach them to dress or brush their teeth. The children were left alone for long periods and were not picked up when they cried or held when they were fed. As a result, the children were considerably underdeveloped in gross motor coordination, fine motor skills, social skills and language development. About 65 per cent of the children three years old and younger exhibited abnormalities in cell and tissue structure and activity resulting from malnutrition.

The prognosis for children deprived of human contact so early and for so long is very poor: scientists have demonstrated that stimulation by adults is critical in the very earliest period of a child's life:

> ... the neurological foundations for rational thinking, problem solving and general reasoning appear to be largely established by age 1... some researchers say the number of words an infant hears each day is the single most important predictor of later intelligence, school success and social competence... the words have to come from an attentive, engaged human being.

Exile from the human family does more than delay or prevent development—it can cause illness, as Maslow also noted, and even death. One study found that in Romania before Ceauşescu's fall, up to 35 per cent of institutionalized children died every year.

But human beings have remarkable resilience if they are drawn back into the circle of care and their crucial needs are satisfied. One orphanage for children with severe disabilities in Babeni, Romania, housed 170 children, all of them deemed "irrecuperable" and therefore cruelly neglected. The institution had no pharmacists, dietitians, psychologists, social workers, physical or occupational therapists, or educational specialists. Although there was adequate food and water, children had such minimal contact with adults that 75 per cent of them didn't know their own name or age and few were toilet trained. Their basic needs for food and water, air and warmth were satisfied, but they lacked human contact—they lacked creative, attentive love. When basic hygiene, bathing, physical therapy, nutritional improvement, greater human contact and psychological counselling were introduced, there was striking improvement within a month.

The physical and psychic consequences of institutional care became obvious when the plight of Romanian children was publicized and started a rash of adoptions. In 1991 alone, 7328 Romanian children were adopted, 2450 by Americans. Of sixty-five American adoptees examined in detail, only ten were judged "physically healthy and developmentally normal," while the rest exhibited "clinical or laboratory findings of serious medical, developmental or behavioural disorders." Scientists found that 53 per cent carried hepatitis B, 33 per cent had intestinal parasites, and many were shorter than normal or exhibited decreased gross motor activity, retarded speech, temper tantrums, gaze aversion or shyness. Clearly, servicing only basic physical needs while failing to provide social contact had profound developmental repercussions. Fortunately, according to scientists, "it appears that many respond well to a loving family environment, improved nutrition and medical and developmental intervention."

But overcoming the trauma of these terrible early experiences isn't always easy. In the 1990s, Americans adopted about nine thousand children from orphanages in Eastern Europe and Russia, and many had serious problems that didn't seem to be easily resolved. According to an article by Sarah Jay in the *New York Times*:

> A child could be hyperactive or aggressive, refuse to make eye contact and have temper tantrums, speech and language problems, attention deficits and extreme sensitivities to touch. The child also might not be able to form emotional bonds.

Victor Groze, who studied 399 families of Romanian adoptees, estimated that a fifth of the children were what he called "resilient rascals" who had overcome their pasts and were thriving, three-fifths were what he called "wounded wonders" who had made vast strides but continued to lag behind their peers, and another fifth were "challenge children" who had shown little improvement and were almost unmanageable.

Years after their adoption, the impact of depriving these children of human touch and meaningful contact was apparent. Once again, the attachment hormones—oxytocin and vasopressin—played a role. A study of Romanian orphans adopted by families in Milwaukee found many of the children exhibited behaviours associated with early neglect, including a lack of attachment to their main caregivers, which was demonstrated by their willingness to seek comfort from unfamiliar adults, even when an adoptive parent was present. The researchers found that children who had experienced early neglect had below-normal levels of vasopressin. As a result of early social deprivation, the release of this critical "cuddle chemical" was inhibited. Sadly, the opportunity to lay the neurological pathways—or, as author Anthony Walsh calls them, the brain's "love trails"—that help secure our relationships had been lost.

Of course, moving to loving, supportive and stable homes went a long way to improving the emotional, psychological and physical health of these children. Love could still have an impact. Children thrived physically when adopted out—their IQs improved as well, as did their ability to express positive emotions—but still, the lack of early contact, bonding and love left an indelible mark.

Educator Lucy LeMare followed the development of 36 Romanian orphans adopted by families in British Columbia. The children had spent anywhere from eight months to four years in orphanages before being adopted. From her work LeMare learned that one of the most important determinants in whether a child would have behavioural difficulties (such as attention deficit or hyperactivity) was the length of time the orphans had been institutionalized. Of the Romanian orphans in LeMare's study group, 43 per cent exhibited these behavioural challenges, compared with only 5 per cent of Canadian-born children and 16 per cent of Romanian orphans who had spent less than four months in an institution. Other studies produced similar results. Although most children from the orphanages showed developmental delays and emotional and behavioural challenges, it was the children who were adopted later (and so spent at least eight months in care) that experienced the greatest challenges.

The security and self-confidence that are provided by a healthy family and that are so necessary for developing a child's self-esteem are amplified by the support of the community in which a family lives. During war, it is difficult for adults to buffer their children against the insecurity that is so damaging to their well-being. In Croatia, civil strife created more than a hundred thousand refugees, many of them displaced children, and often the youngest were separated from their parents. Some 35 per cent of first grade–aged children in camps had been separated from their mothers.

When elementary school children are separated from their families, they exhibit loss of or increase in appetite, sleep disturbances and nightmares, loss of interest in school, difficulties with concentration and memory, irritability, fears, problems in communication, psychosomatic complaints, absence of feeling, rage, unremitting sadness, adjustment disorders or major depression. As one researcher concluded:

> Trusting a close adult is a very important source of support for
> a child. Being aware that parents were not able to protect him
> or her reinforces the traumatic experience of a child exposed to
> the terror of war... the best predictor of positive outcomes for
> the child who survives an intensive stress is the ability of impor-
> tant adults around him, primarily parents, to cope with the trau-
> matic event.

Prehistoric man was, on the whole, a more peaceful, cooperative, un-
warlike, unaggressive creature than we are, and we of the civilized world
have gradually become more and more disoperative, more aggressive and

hostile, and less and less cooperative where it most matters, that is, in human relations. The meaning we have put into the term "savage" is more correctly applicable to ourselves.

—ASHLEY MONTAGU, *The Direction of Human Development*

War is a social, economic and ecological disaster. It is totally unsustainable and must be opposed by all who are concerned about meeting the real needs of all people and future generations. The effect of war is most immediate for those who are killed or maimed or made homeless, but the social and ecological consequences reverberate for generations. Among the children who survive, we don't know the full extent of the psychic damage they have suffered or the degree to which their problems are transmitted to successive generations. War is the ultimate atrocity that dehumanizes victor and vanquished alike; divorcing children from parents, separating families, smashing communities, it deprives its victims of their basic need for love and security in the company of their fellow beings.

Human Communities, Past and Present

Human beings are among the most social of the primates. For 99 per cent of human existence, we have lived in small family groups of nomadic hunter-gatherers. We have depended on family and tribal aggregates for the skills and experience to defend ourselves against predators, marauders and calamity, to capture prey, to gather food, to collect resources for the community. We have gathered together for reaffirmation of place, people and important stages in life, thereby inculcating a sense of belonging, an identity and a worldview. Social groupings have provided other benefits, such as access to mates and long-term relationships and sharing of music, stories, art and recreation.

For most of human existence, we were local, tribal people, perhaps encountering a couple of hundred people and travelling no more than a few hundred kilometres in a lifetime. We didn't have to worry about what tribes on the other side of a lake, mountain or ocean were doing. Nature seemed vast and endlessly self-renewing, and if we were able to degrade resources with fire or tools, we would move elsewhere. But now we have moved across the planet, and the collective impact of all humans on Earth is reverberating throughout the biosphere. We have to consider the sum of the effects that our species have created through fishing, logging, polluting, damming, developing and so on, and it's not easy because we have never operated that way. So, negotiations such as the ones that took place at the Earth Summit in Rio and the climate conferences in Kyoto and Paris were frustratingly difficult and slow.

Throughout history, people have depended heavily on their main survival trait—a curious and inventive mind endowed with a prodigious memory and forethought. From earliest times, people assessed the potential consequences of their actions based on what they already knew from past experiences. Like a chess player planning moves, their minds flashed back and forth from past to future to present, as no other species had ever been able to do.

Essentially, the early humans were carrying out what would now be called a "cost-benefit" analysis, weighing the potential benefits against the long-term costs. Accumulated tribal knowledge was brought to bear on important decisions through elaborate ceremony and ritual, which reinforced the ties between members of the community. These days we give lip service to carrying out cost-benefit analyses, but increasingly, as the consequences of what we do collectively as a species have become more far-reaching, the repercussions can no longer be predicted. And the definitions of "cost" and "benefit" have changed. Whereas in the past the most important factor was the long-term survival or well-being of the family or group, today decisions are made based on the implications for a company, job, market share or profit. So

Animal Friends

Human societies are not the only groups that forge bonds and use strategies such as touch therapy to support their community. Other animals, particularly primates, show complex social behaviours that strengthen relationships. Baboons, for instance, show their affection and affirm social bonds through grooming. When they groom, levels of stress hormones, the glucocorticoids, decline. Like humans, baboons turn to one another in times of stress. After a death in their community, they broaden and strengthen their grooming networks, reaching out to others for comfort. As they groom, the levels of stress hormones, which increased with the trauma, begin to fall and eventually normalize. Chimpanzees exhibit similar behaviour and will console a victim after a fight, with hugs, pats on the back or bouts of grooming.

Many non-primate species show sympathy toward one another as well. By paying attention to the needs of individuals and reinforcing social bonds, they strengthen the community. Sperm whales will surround an injured member of their pod, for instance, and false killer whales have stayed with a critically injured member of the pod for days, moving on only when the individual died. Most animals pay only a passing interest in dead members of their own species, but elephants exhibit what could be construed as mourning. They pay particular attention to the skulls, ivory and bones of dead elephants, even those long deceased. Elephants have been seen gently rocking pieces of ivory back and forth with their feet and will also pick up ivory and carry it around in their trunks.

Play is another way to bond with family and community. Wild apes, for instance, may use humour to strengthen social bonds and to ease the transition when new individuals join their group. Like humans, orangutans, gorillas and chimps laugh when tickled and use smiles and vocalizations that seem to suggest play and humour. □

we assess costs and benefits within a very different framework of values, ignoring, for example, the health of the community or ecosystem. We have gotten out of the habit of thinking about the things that really matter to us, or perhaps we have a perverted sense of what really matters. Instead of thinking of different bits and pieces of our lives, we must think in a more complete way. John Robinson and Caroline Van Bers have suggested some ways to get a more complete picture of our well-being:

> Although we may struggle with the concept of ecological sustainability, all humans can understand the idea of social well-being. We are able to assess our personal well-being without much trouble. It depends on how we feel about our prosperity, the place we live, our families and friends, our physical health, and a range of other conditions. Similarly, we can usually assess the well-being of our neighbourhood or our community using indicators that are almost intuitive—community vitality, absence of conflict, healthy trees and streams, and so on…

The need for community and its rituals is an ancient need. It has been built into the human psyche over thousands of generations and hundreds of thousands of years. If it is frustrated, we feel "alienated" and fall prey to psychiatric and psychosomatic ills.
—ANTHONY STEVENS, "A Basic Need"

The threats to community are from more than warfare; we are in the grip of a mindset called modernity that views whatever is modern and recent as the best. Conversely, in this mindset, whatever is ancient or traditional is seen as primitive and less desirable. Overwhelmed by the incredible changes brought about by technology and materialism, we have accepted a widespread belief that somehow human beings today are different from people in the past; because we have more information, have travelled more and are better educated than all our predecessors, surely our thoughts and needs are more sophisticated, on a different plane from those of all preceding generations. As we cut the ties to our ancestors we find ourselves helpless in a world of dizzying change, cut off from memory and forethought. Without context, information becomes meaningless; without perspective, events cannot be evaluated; without connections in time and space, we are lonely and lost.

The placenta must be buried with ceremony in the compound with the witch-doctor present. As the navel cord ties an unborn child to the womb, so does the buried cord tie the child to the land, to the sacred Earth of the tribe, to the Great Mother Earth. If the child ever leaves the place, he will come home again because the tug of this cord will always pull him toward his own.

When I go home... I shall speak these words: "My belly is this day reunited with the belly of my Great Mother, Earth!"
—PRINCE MODUPE, *I Was a Savage*

Rituals are a public affirmation of meaning, value, connection. They tie people to each other, to their ancestors and to their place in the world together. Anthony Stevens points out that while industrialized nations have achieved extraordinary levels of wealth, consumer goods and amenities, more and more people yearn for community and rituals that bind them together. This desire intensifies as family breakdown becomes more common, and adults and children alike suffer the consequences.

FOR THOUSANDS OF years, small communities of people ensured relative tranquility while providing for the social needs of their members. The explosive rate at which our species has been converted to an urban creature has been accompanied by a deterioration of the social fabric that held people together. The twentieth century witnessed an unprecedented shift from predominantly rural community living to big city living. In cities, distanced from nature and the primary means of production like agriculture, fishing, logging and even manufacturing, we accept that it is the economy that provides our needs. Technology has enabled us to travel rapidly and communicate over vast distances, while tele-vision, computers and portable entertainment devices sever the shared activities with neighbours and communities. Consumerism has taken the place of citizenship as the chief way we contribute to the health of our society. Economic rather than social goals drive government and corporate policies. The resulting high levels of unemployment produce stress, illness, and family and community breakdown. Stable communities and neighbourhoods are a prerequisite for happiness, for productive and rewarding lives, for a crucial sense of security and belonging. They are a bottom line for the health and happiness of human beings. It is not economics that creates community but love, compassion and co-operation. Those qualities exist in individuals and are expressed between people. And they cannot be fully expressed in isolation, without context, cut off from their place in time and space, their source in the natural world.

The stability of family—whatever its form—within a community provides an environment within which a child develops curiosity, responsibility and inventiveness. Ecological degradation—deforestation, topsoil loss, pollution, climate change and so on—destabilizes society by eroding the underpinnings of sustainability. This consequence was graphically illustrated in 1992, when all commercial fishing of northern cod in the Canadian province of Newfoundland was suspended. Overnight, forty thousand jobs were lost as the foundation of that society for five centuries vanished. All across Canada, towns boomed as forests were clearcut around them, only to crash when the trees were gone. The coast of British Columbia is dotted with villages that once supported fishing fleets and canneries but were abandoned as salmon populations declined. Ecological health is essential for full community health.

War, terrorism, discrimination, injustice and poverty mitigate against that social stability that is so important. Chronically high levels of unemployment, such as those found in the Atlantic provinces in Canada, on American Indian reservations or in Australian Aboriginal communities today, result in despair, alcoholism, illness, even death. The need for meaningful employment is critical to the well-being not only of family but of community. Besides the economic benefits to government and individuals, there are compelling reasons to seek full employment as a social goal.

An economy was once created to serve people and their communities. Today economic rationalists contend that people must sacrifice and give up social services for the economy. As we reflect on our fundamental needs as social animals, it is clear that families and communities assured of biodiversity, full employment, justice and security constitute the real non-negotiable starting point in the delineation of a sustainable future.

The law of love will work, just as the law of gravitation will work, whether we accept it or not... a man who applies the law of love with scientific precision can work great wonders... The men who discovered for us the law of love were greater scientists than any of our modern scientists... The more I work at this law, the more I feel the delight in life, the delight in the scheme of this universe. It gives me a peace and a meaning of the mysteries of nature that I have no power to describe.

—MAHATMA GANDHI, quoted in P. Crean and P. Kome, eds.,
Peace, a Dream Unfolding

From family to neighbourhood, from neighbourhood to nation, out into the commonwealth of our species—the connection seems to attenuate as it becomes more inclusive. But as we explore the continuum of relationships in any human life we start to see that the circle of inclusion extends further still; the "continent" of which we are each a part encompasses the Earth.

The "law of love" is as fundamental, and as universal, as any other physical law. It is written everywhere we look, and it maps our intimate connection with the rest of the living world.

Biophilia: Reviving Our Evolutionary Links

Think of where we have been all this time on Earth. For almost our entire existence we have lived completely immersed in the natural world, dependent on it for every aspect of our existence. Moving through the landscape around us, led by the seasons, we lived lightly on the land and were sustained by its biological plenitude.

If the richness of plant and animal diversity in today's Africa is any indication of past abundance (and fossil remains corroborate this abundance), our original ancestors were surrounded by a staggering array of animal and plant forms. We emerged from and were completely bound up in the matrix of life forms that shared our surroundings. They were more than just our genetic kin and potential prey; they were our *companions*, sharing the clear night skies and constantly announcing their presence with their calls. To this day hunter-gatherer societies treat the sources of their food with respect and sympathy: !Kung hunters in the Kalahari fast before the hunt to make themselves worthy of their task. After a kill they thank the animal for the gift of life and then carry the carcass back to the camp for a ritual sharing. Food was a gift of the flesh of other beings that had to be properly acknowledged. According to Ivaluardjuk, an Inuit man:

> The greatest peril of life lies in the fact that human food consists
> entirely of souls. All the creatures that we have to kill and eat, all
> those that we have to strike down and destroy to make clothes
> for ourselves, have souls like we have, souls that do not perish

with the body, and which must therefore be propitiated lest they should revenge themselves on us for taking away their bodies.

The evolutionary context of human history makes it plausible that the human genome—the DNA blueprint that makes us what we are—has over time acquired a genetically programmed need to be in the company of other species. Edward O. Wilson coined the term "biophilia" (based on the Greek words for "life" and "love") for this need. He defined biophilia as "the innate tendency to focus on life and life-like processes." It leads to an "emotional affiliation of human beings to other living things... Multiple strands of emotional response are woven into symbols composing a large part of culture."

Elders, poets and philosophers in all cultures, including our own, have expressed a similar sense of brotherhood or sisterhood, of mutual compassion and common interest with the rest of the living world—a relationship that can only be described as love. Its source is "fellow-feeling": the knowledge that we are, like all other forms of life, children of the Earth, members of the same family.

The indescribable innocence and beneficence of Nature—of sun and wind and rain, of summer and winter—such health, such cheer, they afford forever!... Shall I not have intelligence with the Earth? Am I not partly leaves and vegetable mould myself?
—HENRY DAVID THOREAU, *Walden*

In urban environments, our genetically programmed need to be with other species is usually thwarted, leaving us yearning. These days, biophilia has to be satisfied with sadly diminished opportunities—gardening, pets, visits to zoos. It is not an accident, Wilson says, that more people visit zoos than attend all major sports events combined. The pull of biophilia can be so compelling that even something as simple as a room with a view seems to make a real difference. At a state prison in southern Michigan, for instance, prisoners whose rooms had windows facing farmland and forest had 24 per cent fewer medical visits than those prisoners with rooms facing an interior courtyard.

Meaningful interaction with other living things can help us heal. In one study, researchers followed seventy-one new pet owners and compared them with petless people. Within a month, pet owners showed a decrease in health problems. Pets, particularly dogs, are used in many therapy programs to provide comfort and companionship to people in hospitals, nursing homes, schools and community centres.

We can find healing in the garden, too. Mental well-being is enhanced by horticultural therapy, and gardens are becoming an integral part of the healing therapies at schools, nursing homes, hospitals, prisons and more. Writer Oliver Sacks recounts the importance of a garden to his personal healing after a serious leg injury, when, after spending almost a month in a room without a view, he was taken into the garden:

> This was a great joy—to be out in the air... A pure and intense joy, a blessing, to feel the sun on my face and the wind in my hair, to hear birds, to see, touch, and fondle the living plants. Some essential connection and communion with nature was re-established after the horrible isolation and alienation I had known. Some part of me came alive, when I was taken into the garden...

Clearly, all of these examples show that interactions with living things do make a difference, but it is also critically important that we engage with the wild in the world as well. Wilderness experiences are not solely the domain of adventure travel and high-adrenalin sports—something as simple as a walk in a local park or a rest on the banks of a creek can go a long way to provide respite, restoration and reconnection. For example, I have attended meditation sessions for terminal cancer patients, people who have ridden a roller coaster of hope and despair after chemotherapy, radiation and surgery. They attested to the healing and soothing effects of nature. They told me that living with their illness allowed them to "truly live" for the first time, and almost all of them made reference to the importance of "being in nature," whether walking in the woods, strolling a beach or resting on a farm or at the cottage.

The truth is that we have never conquered the world, never understood it; we only think we have control. We do not even know why we respond in a certain way to other organisms, and need them in diverse ways so deeply.
—EDWARD O. WILSON, *Biophilia*

Watch children respond to a wasp or butterfly. Infants seem drawn to an insect's movement and colour, often reaching out to touch it. They exhibit neither fear nor disgust, only fascination. Yet by the time they enter kindergarten, enchantment with nature has often been replaced with revulsion as many children recoil in fear or loathing at the sight of a beetle or fly. By teaching children to fear nature, we increase our estrangement and fail to satisfy our inborn biophilic needs. We sever the connections, the love that infuses our actions with compassion for our fellow beings. It is sad that extreme crises such as a nervous breakdown, a severe injury, loneliness or death are needed to bring us back to the healing comfort of home.

Biophilia provides us with a conceptual framework through which human behaviour can be examined and evolutionary mechanisms suggested. It is a new story, which includes us in the living world around us, restoring us to our long-lost family. Studies that support the biophilia hypothesis are accumulating. For example, architecture professor Roger S. Ulrich reports that a consistent finding in well over 100 studies of recreation experiences in wilderness and urban natural areas has been that stress mitigation is one of the most important verbally expressed perceived benefits.

It appears to be scientifically verifiable that human beings have a profound need for an intimate bond with the natural world, leading to the suggestion that

> the degradation of this human dependence on nature brings the increased likelihood of a deprived and diminished existence... Much of the human search for a coherent and fulfilling existence is intimately dependent upon our relationship to nature.

Love, as we know from our own experience, is formative: it shapes the giver as it affects the receiver, because it forms a union between the two. The beloved child learns from that experience that he or she is lovable, enabling the child to love others. Biophilia teaches us the same lesson, according to Wilson:

> The more we know of other forms of life, the more we enjoy and respect ourselves... Humanity is exalted not because we are so far above other living creatures but because knowing them well elevates the very concept of life.

Wedding Psychology and Ecology

Psychologists have generally emulated reductionist science, concentrating on an individual psyche with little regard to the environment in which a person exists. Reductionism focuses on isolating, separating and controlling fragments of nature and has been a powerful way of knowing, providing profound insights into the properties and behaviour of that bit of nature. But in focusing, we lose sight of the context, rhythms, patterns and cycles within which that part belongs. It is thus a terribly limited way of knowing. Of course, relationships with other human beings are a critical part of psychoanalysis, but the other biological creatures in our surroundings, the chemical and physical circumstances where we live, work and play, also affect us immensely. Ecopsychology is a long-overdue attempt to widen our scope to include environmental determinants of our psychic makeup and health. When we forget that we are embedded in the natural world, we also forget that what we do to our surroundings we are doing to ourselves. Ecopsychology is an attempt to reconnect us with our natural home and to remedy some of the harm caused by our exile in the modern city. Ecopsychologists argue that because we are separated from nature we risk our emotional health. Instead of trying to adjust to the existing social order and accept the status quo, they argue that for true mental health we must challenge the norm and take into account that the needs of both Earth and humans are interdependent and interconnected. As Anita Barrows has pointed out:

> It is only by a construct of the Western mind that we believe ourselves living in an "inside" bounded by our own skin, with everyone and everything else on the outside. The place where transitional phenomena occur... might be understood in this new paradigm of the self, to be the permeable membrane that suggests or delineates but does not divide us from the medium in which we exist.

If we continue to think of ourselves as separate from our surroundings, we will not be sensitive to the consequences of what we are doing, so we can't see that our path is potentially suicidal. If we do not see ourselves as part of the natural world and become further detached, we risk a greater sense of loneliness, a lack of meaning, purpose and sense of belonging. Without experiences in nature, we develop ignorance and apathy. Thus, for example, our eyes and noses may inform us that city air is no longer the colourless, odourless, tasteless, invisible gas defined in physics texts, yet we seem unaware that there may be a link between air pollution and the rising number of children with asthma.

Our schism from nature is reinforced by the way we construct our habitat. Most of humanity in the industrialized world and a rapidly increasing number in the developing nations live in cities where town planners, architects and engineers dictate the nature of our surroundings.

In the urban environment that is today's most common human habitat, science and technology perpetuate the illusion of dominance and shape the way we see the world. Cities manifest a way of thinking that reflects mechanical or technological models based on standardization, simplicity, linearity, predictability, efficiency and production. As Vine Deloria points out, this mindset is mirrored in the kind of habitat we have created:

> Wilderness transformed into city streets, subways, giant buildings, and factories resulted in the complete substitution of the real world for the artificial world of the urban man... Surrounded by an artificial universe when the warning signals are not the shape of the sky, the cry of the animals, the changing of the seasons, but the simple flashing of the traffic light and the wail of the ambulance and police car, urban people have no idea what the natural universe is like.

The place where we spend most of our lives moulds our priorities and the way we perceive our surroundings. A human-engineered habitat of asphalt, concrete and glass reinforces our belief that we lie outside of and above nature, immune from uncertainty and the unexpected of the wild. We can see how much life and values have changed from our origins by looking at the remaining pockets of Indigenous people who still manage to live traditionally as our ancestors did for most of human existence. In the words of Paul Shepard:

> Their way of life is the one to which our ontogeny has been fitted by natural selection, fostering cooperation, leadership, a calendar of mental growth, and the study of a mysterious and beautiful world where the clues to the meaning of life were embodied in natural things, where everyday life was inextricable from spiritual significance and encounter, and where the members of the group celebrated individual stages and passages as ritual participation in the first creation.

To Shepard, the fundamental human relationship that shapes the person-to-be remains the child-mother bond. But once we are secure in that bond, our surroundings influence us powerfully.

Shepard also suggests that unless we actually experience nature at very specific intervals in childhood, we fail to trigger the emotional bonding with the wild world that affects the way we treat it as adults. Consequently, we lack the constraints against developing materialist, nihilist and other ecologically destructive attitudes, and we become

> careless of waste, wallowing in refuse, exterminating enemies, having everything now and new, despising old age, denying human natural history, fabricating pseudotraditions, being swamped in the repeated images of American history. They are the signs of private nightmares of incoherence and disorder in broken climates where technologies in pursuit of mastery create ever-worsening problems—private nightmares expanded to a social level.

Clearly, we need a change in direction that leads us to a greater awareness of our connection to—indeed, our inseparability from—nature. As delegates at a 1990 psychology conference at Harvard concluded, "If the self is expanded to include the natural world, behavior leading to destruction of this world will be experienced as self-destruction."

To heal our planet and ourselves, we need to move away from "pathological individualism," as ecopsychologist Sarah Conn calls it. After all, she explains, "we don't live on the Earth; we live in it." We must begin to see ourselves as part of our environment, not separate from it. Our identity goes beyond flesh, blood and our actions and thoughts. Our identity includes our natural world, how we move through it, how we interact with it and how it sustains us. If those connections of belonging, to the Earth and to each other, are not there, we must forge them, reconnecting people and nature so that they both can heal.

Love Makes Us Human

Built into the fundamental properties of matter is mutual attraction that could be thought of as the basis of love. For human beings, love, beginning with the bond between mother and infant, is the humanizing force that confers health in body and mind. Receiving love releases the capacity for love and compassion that is a critical part of living together as social beings. That love extends beyond those of our own species—we have an innate affinity for other life forms. If we are to deliberately plot a sustainable future, the opportunity for each of us to experience love, family and other species must be a fundamental component.

8

Sacred Matter

Soul clap its hands and sing, and louder sing
For every tatter in its mortal dress.
—W.B. YEATS, "Sailing to Byzantium"

MEETING BASIC, INALIENABLE physical needs is just the beginning of human well-being. As we have seen, denial of love, companionship and community causes serious, sometimes fatal damage to a developing human being. But beyond physical and social needs, we have yet another need, one that is just as vital to our long-term health and happiness. It is a need that encompasses all the rest, an aspect of human life that is so mysterious it is often disregarded or denied. Like air and water, like the love and companionship of our kind, we need spiritual connection; we need to understand where we belong.

Our stories tell us where we come from and why we are here. In the beginning, these stories say, there was water, and then there was sky and fire, there was Earth, and there was life. We humans crawled out of the womb of the planet, or we were shaped out of clay and water, carved from twigs, compounded of seeds and ashes, or hatched from the cosmic egg. One way or another, we were made from the sacred elements that together compose the Earth. We are made from the Earth, we breathe it in with every breath we take, we drink it and eat it, and we share the same spark that animates the whole planet. Our stories tell us this, and so does our science.

According to our myths we were made for a variety of purposes. We are commanded to be fruitful and multiply, like all other living things; to rejoice and give thanks to the creator; to name and care for the wonders of creation, or simply to give voice to them. Spider Woman in the Hopi myth says to Sotuknang, the manifest god: "As you have commanded me I have created these First People. They are fully and firmly formed; they are properly coloured; they have life; they have movement. But they cannot talk. That is the proper thing they lack. So I want you to give them speech. Also the wisdom and power to reproduce, so that they may enjoy their life, and give thanks to the Creator."

Creation stories create, or re-create, the world human beings live in, shape what we see and suggest the rules by which we should live. Unbelievably numerous and diverse, these tales of the Beginning of Everything are considered by the peoples who live by them the most sacred of all the stories, the origin of all the others. Myths help us to reconcile conflicts and contradictions and describe a coherent reality. They make a meaning that holds the group together and express a set of beliefs; even in our skeptical society, we live by myths that lie so deep we believe them to be reality.

As well as telling us where we come from, our myths also tell us that something went badly wrong, that we humans have been exiled from home, ousted from the garden. Many different stories describe how we lost our place in the harmony of creation. The first man and the first woman ate the fruit of the tree of the knowledge of good and evil, believing it would make them like gods. Prometheus stole the sacred fire that was reserved for the gods, and drew down punishment on men. Many African myths tell a similar tale: "All the animals watched to see what the people would do. They made fire. They rubbed two sticks together in a special way and thus made fire. The fire caught in the bush and roared through the forest and the animals had to run to escape the flames." In this story from the Yao in northern Mozambique humans brought fire and killing to "the decent peaceful beasts." This cruelty drove the gods themselves from the face of the Earth. Most belief systems include such stories, describing how we disobeyed the gods, tricked them, tried to be like them, flouted heaven. Acting differently from the rest of creation, separating ourselves from divine will, we broke the harmony. Because the story of our fall is common to most cultures, the problem must be human, not cultural. We live in a world where things go wrong, where conflict and tragedy are common, where we are often lonely and confused, and our myths give us reasons for this disorder.

What makes us so different from other creatures on Earth? Disobedience, quarrelsomeness, ambition, greed—these are the crimes, we tell ourselves, that have set us apart from the rest of nature. But they may be the consequences of being conscious. Consciousness and its creation, culture, are the primary adaptive tools of human beings. Our giant brain allows us to see patterns by discerning repetition, similarity and difference. From this we gain history and we gain foresight—we can plan. Because we can learn from experience, we can teach our children more than we knew when we were their age. We can change more rapidly than evolution would allow us to, responding to threats by drawing from our experience and deciding to alter the way we live.

> *Your shadow at morning striding behind you*
> *Or your shadow at evening rising to meet you;*
> *I will show you fear in a handful of dust.*
> —T.S. ELIOT, *The Waste Land*

Consciousness helps us make sense of the sensory stimuli that bombard us every waking moment. Without consciousness, speech, for instance, would simply be a physical phenomenon—sounds created by waves of energy. Instead, through these sounds, we intuit emotion and relationships. Sounds facilitate self-awareness: we open *our* mouths, and sounds emerge. Sound may be physics, but our awareness and interpretation of these sounds—through our consciousness—give personal context to these phenomena and to our lives.

With consciousness, we are able to perceive that there is a relationship between our environment and ourselves. The warmth of the sun makes you happy; you see the vivid orange of a poppy, you smell burning toast; you can feel pain and inflict it. Yet each of these experiences is subjective, and this variation of perception among individuals is what makes consciousness so difficult to grapple with. Do individuals see the same "quality" of blue? Do their internal thoughts take the same pathways? How do they feel pleasure and pain? With consciousness, we become aware of the mental lives of others. And this awareness, in turn, gives rise to more abstract qualities such as compassion, guilt and envy.

Once on the scientific fringe because of its abstract and subjective nature, the study of consciousness is growing as neuroscientists, psychologists and philosophers try to create an understanding of how consciousness works. As philosopher David Chalmers says, "There is nothing that we know more intimately than conscious experience, but there is nothing that is harder to explain." Chalmers posits that there are different types of problems in studying consciousness: the easy problems, such as how physical processes in our body give rise to cognition (in essence, the objective qualities; the things we can measure through cause and effect), and the hard problems, such as the nature of our thoughts, perceptions and emotions. What physical processes in our brains give rise to these subjective experiences? Herein lies the real mystery to the mind. How are the chemical and electrical changes in our brains translated into our thoughts, emotions and behaviours? Perhaps we will never know the answer. Chalmers, for example, thinks that maybe consciousness is like space, time and mass—physical realities of our universe that just are.

As we begin to understand consciousness in ourselves, we can begin to extend this understanding to answer questions such as when consciousness in humans begins or whether non-human animals have consciousness as well. Some studies show that animals other than humans show signs of consciousness, or at least self-awareness, the ability to recognize one's own existence. The most common test of self-awareness is to see whether subjects see themselves or what they perceive to be a stranger when they look into a mirror. A chimpanzee, for instance, will wipe a spot of paint off its forehead when it looks in a mirror, thus demonstrating an understanding that the paint is on its own forehead and not on that of another chimpanzee. Chimpanzees share this ability to *become* self-aware (they usually do not recognize themselves instantaneously) with orangutans and perhaps even dolphins. So what does self-recognition mean? By being aware of our own self and how our experiences affect us, we can use this knowledge to infer an understanding of the mental state of others. We learn that we can be altruistic or cruel, empathetic or selfish. With consciousness we forge our place in the human community.

Consciousness brings layers of subjective qualities to our lives, but it has its drawbacks. Conscious of time, we know our origin and our destiny: we know that we are doomed to die. That knowledge is always with us: the precious *I, me, myself,* the centre of each consciousness, will eventually disappear.

I lift my voice in wailing. I am afflicted, as I remember that we must leave the beautiful flowers, the noble songs; let us enjoy ourselves for a while, let us sing, for we must depart forever, we are to be destroyed in our dwelling place.

It is indeed known to our friends how it pains and angers me that never again can they be born, never again be young on this Earth.

Yet a little while with them here, then nevermore shall I be with them; nevermore enjoy them, nevermore know them.

Where shall my soul dwell? Where is my home? Where shall be my house? I am miserable on Earth.

We take, we unwind the jewels, the blue flowers are woven over the yellow ones, that we may give them to the children.

Let my soul be draped in various flowers; let it be intoxicated by them; for soon must I weeping go before the face of our Mother.

—AZTEC LAMENTATION, in Margot Astrov, ed.,

American Indian Prose and Poetry

Some deaths we understand to be temporary. As Earth rolls around the sun the seasons roll with it, and humans have learned from them that nature's death precedes rebirth. Perhaps the Eden of our myths was the tropical region in which we first evolved, where the sun never slips down the sky in winter, the air is always warm and humid, and the trees bear fruit in a measured endless sequence. Since we dispersed into temperate regions and beyond we have watched the leaves fall, endured the grip of winter and learned that spring will come again, with the appropriate rituals and sacrifices. But we have also seen age darken into death, watched children struck down, endured the loss of beloved individuals. Unlike the deaths of nature, our loss is permanent. As Shelley expresses it in his elegy *Adonais*, "Ah woe is me! Winter is come and gone, / But grief returns with the revolving year." Time in nature is cyclical or cumulative; human time is linear. Because of this contradiction between the recurrence of nature and the finality of our own fate we reach for something eternal, something absolute, unchanging, outside time: the essential me-ness of me, the soul, the spirit. Without water, air, energy, food, without other forms of life and other human beings, we die. But we are just as crucially dependent on the idea of spirit. Without that we are truly doomed, drowned in time and change, forced to watch the gap between *now* and *the end* as it inexorably shrinks. Friends, family, all the joys and beauty of life are threatened by time and death, and we need spirit to heal that sorrowful knowledge.

"Spirit" is a powerful, mysterious word; in English its meanings spread like an invisible web through every level of existence. It is air, as we saw in an earlier chapter, it is breath, and by extension it is life and it is speech. It is the power of divine creation, moving over the waters, and it is divinity itself—the Great Spirit, the Holy Spirit, the Lord of All. Spirits are volatile, invisible, powerful, and some are eternal. They may intoxicate, invigorate, inhabit, haunt, or they may express the essence of something. Above all, they animate the world—make it holy. "Spirituality," as we conceive it, is the apprehension of the sacred, the holy, the divine. In our modern world we see matter and spirit as antithetical, but our myths reveal a different understanding. They describe a world permeated by spirit, where matter and spirit are simply different aspects of the totality: together they constitute "being." All cultures have believed in power beyond human power, in life beyond death, in spirit. Many have believed in an animated, inhabited, sacred world surrounding them, the natural world that constitutes reality. These beliefs restore our sense of belonging, of being-with, which is threatened by our dividing, conquering brain; they provide us with rules and rituals for restoring the harmony, for re-entering and celebrating the world we are part of. The mythmaking of our mind, its ability to find coherence in chaos, to create meaning, may be our species' antidote to the risks of consciousness—a cure for death.

Spirit Worlds

In Cree, we don't have animate-inanimate comparisons between things. Animals have souls that are equal to ours. Rocks have souls, trees have souls. Trees are "who," not "what."
—TOMSON HIGHWAY, quoted in Suzanne Methot,
"The Universe of Tomson Highway"

Traditional cultures live in an animated world. Mountains, forests, rivers, lakes, winds and the sun may all have their presiding deities, while each tree, stone and animal may have, or be, a spirit. The spirits of the dead, or of the unborn, may also be eternally present, acting powerfully in the living world, part of the endless circle of time. Such worldviews may see all death, including that of humans, as simply one stage in the continuum of birth, life, death and rebirth that we see in nature. Human beings are included in this totality of creation, participating in various ways in the creative mind of the living Earth. Instead of being separated from the world because of their unique consciousness, they belong to a conscious world in which everything interacts with everything else in a process of continual creation. Contained within this worldview are the rituals that allow wrongs to be righted, spirits to be propitiated, the world to unfold as it ought. These rituals are the responsibility of the human part of creation (perhaps because the disruptions are often caused by us).

The traditional worldview of Hawaiians is just one of innumerable examples of this kind of worldview. Michael Kioni Dudley summarizes a portion of these beliefs: "Hawaiians traditionally have viewed the entire world as being alive in the same way that humans are alive. They have thought of *all* of nature as conscious—able to know and act—and able to interrelate with humans... Hawaiians also viewed the land, the sky, the sea, and all the other species of nature preceding them as family—as conscious ancestral beings who had evolved earlier on the evolutionary ladder, who cared for and protected humans, and who deserved similar treatment (*aloha'aina* [love for the land]) in return."

Similarly, Australian Aborigines live in a land that is constantly created, partly by their agency. The Ancestors sang the world into existence in the Dream Time. Now, as David Kinsley describes, contemporary Aborigines "have the responsibility of perpetuating the sacred character of the land by re-creating it, or remembering it," using the same songs, handed down through the generations. An Aboriginal woman becomes pregnant when she passes a sacred part of the landscape and a spirit ancestor decides to enter her. The spirit grows to term and is born into the world as a human being. "That is, every human being is in some essential way a spirit of the land, a being who has an eternal, intimate connection with the land. He or she is an incarnate spirit of the land, living temporarily in human shape and form." To know that person's true identity, Aborigines have to discover from which sacred site the spirit ancestor came. "A person is not simply the offspring of his or her physical parents. Each individual is primarily an incarnation of the land, a spirit being who belongs intimately and specifically to the local geography. Their beliefs concerning spirit impregnation are an unambiguous statement that human beings are rooted to the land and will become disoriented, suffer, and ultimately die if uprooted and transported outside the location of their birth."

The sense of the spirit, of the supernatural, is shared by a vast number of world religions. But why did religions originate at all? Some think that religion provided a framework for interpreting the phenomena observed by early humans—the motion of the sun, the passing of the seasons—and answered questions regarding who or what controlled these phenomena, and, perhaps most important, what happens after death. As new societies tried to answer these questions, religious belief systems emerged, each with its own rituals, leaders, expectations, moral codes and answers to life's big questions.

Religion may have also been an evolutionary advantage. By bringing people together as members of a community, with shared ideas about the world, religions provided the social glue that bonded societies. Some speculate that by offering a structure for passing on knowledge about the world, religions provided an edge that helped people survive.

By giving answers about our destiny once we ceased to be mortal, religion also gave context and meaning to lives that were fleeting and often rife with danger and chaos. Religion might have served to ease people's emotional burdens and lessen their fears. Regardless of whether life hereafter would promise reincarnation or heaven, somehow just "knowing" their fate might have provided some comfort. (Today, we know that religion can deliver a "feel-good factor." For instance, some studies show that compared with non-religious people, people who are religious live longer and healthier lives.)

There are dozens of religions in the world, mainstream or otherwise, yet, interestingly, through virtually all of them weaves a common thread—they share stories of the supernatural, of souls or miracles or divine creation, stories that seem to contradict what science tells us about the world. These spirited beings can part water and walk through walls; they can inhabit the rocks and the trees, even the heavens. Fanciful? Perhaps, yet even those most grounded in the physics of our universe can be transfixed by these compelling ideas. Psychologist Paul Bloom feels that there is a good reason for this inclination—he thinks we are predisposed to believe in the spirit. Bloom proposes that we are natural dualists—inclined to believe, particularly when we are young, that our physical body and our conscious entity (a mind or soul) are distinct. For many people, Bloom suggests, it doesn't matter that this idea conflicts with science; it just "feels right." For instance, many people—even those who are not particularly religious—seem to talk quite naturally about people's souls going elsewhere after death. Even if we don't believe it, we can *understand* the idea of a body without a soul (a corpse) and even, perhaps, a soul without a body (angels, spirits, a deity, such as God, who listens to our prayers).

Children in particular tend to intuitively believe in the supernatural and to believe that bodies and souls are distinct. In one study, for instance, children were told a story of a dying mouse. When questioned, the children understood that the mouse's body was gone and that it couldn't hear or go to the bathroom, but more than half of the children felt that the dead mouse would still like cheese and love its mother. In their young minds, the mind and body of the mouse were distinct. Its body might be gone, but its soul had survived. With this example and many others, Bloom suggests that the notion of life after death is something we are naturally wired to consider rather than something that is taught through formal religion. As Bloom says:

Religious teachings certainly may shape the specific beliefs we hold; nobody is born with the idea that the birthplace of humanity was the Garden of Eden, or that the soul enters the body at the moment of conception, or that martyrs will be rewarded with sexual access to scores of virgins. These ideas are learned. But the universal themes of religion are not learned. They emerge as an accidental by-product of our mental systems. They are part of human nature.

All religions explore the place of people in the natural and social worlds around them. They provide explanations for mysteries such as death and disorder, and use myths and moral teachings to relate human and non-human spheres. The earliest forms of contemporary world religions, such as Hinduism, Judaism, Christianity and Islam, presented an animated, integrated world similar to that of traditional worldviews. As Lao Tzu puts it in the *Tao Te Ching*:

> The virtue of the universe is wholeness,
> It regards all things as equal.

But some of those world religions have shifted ground over the past centuries, supporting the development of a very different picture of reality and our place in it.

Alienation of the Spirit

Now we are no longer primitive; now the whole world seems not-holy. We have drained the light from the boughs in the sacred grove and snuffed it in the high places and along the banks of the sacred streams. We as a people have moved from pantheism to pan-atheism.
—ANNIE DILLARD, *Teaching a Stone to Talk*

Here in the West we have exorcised the spirits and cut ourselves loose from the living web of the world. Instead of seeing ourselves as physically and spiritually connected to family, clan and land, we now live chiefly by the mind, as separate individuals acting on and relating to other separate individuals and on a lifeless, dumb world beyond the body. Applying our mind to the matter around us, we have produced an extraordinary material culture: cities and highways and toasters and blenders, computer technology, medical technology, paper clips, assault rifles and television sets. But we find ourselves separated, fragmented, lonely, fearful of death. We have coined a word for this state of mind: "alienation," which means being estranged. We are strangers in the world; we no longer belong. Because it is separated from us we can act on it, abstract from it, use it, take it apart; we can wreck it, because it is *another*, it is *alien*. We may feel despair, grief and guilt about the damage we cause—but we seem unable to change the way we live. How has this happened? Is it because we have lost our religion? Or is that a consequence rather than a cause? Perhaps it is the inevitable consequence of "modernization," as human societies have moved away from immediate dependence upon the land.

That movement away from the natural world was made possible by a quite remarkable train of thought—the ideas that shaped our civilization. Today we take those ideas so much for granted that we see them not as ideas (which can be rethought, revised, discarded) but as reality. Many thinkers trace the origins of our particular and violent fall from grace, our exile from the garden, back to Plato and Aristotle, who began a powerful process of separating the world-as-abstract-principle from the world-as-experience—dividing mind, that is, from body, and human beings from the world they inhabit. In the process they laid the groundwork for experimental science.

Through Galileo, who identified the language of nature as mathematics (an abstract language invented by humans), and Descartes, who learned to speak that language powerfully, the modern world emerges. Descartes's famous definition of existence ("I think, therefore I am") completes a new myth about our relationship to the world: human beings are the things that think (the *only* things, and that is *all* they are), and the rest of the world is made up of things that can be measured (or "thought about"). Subject or object, mind or body, matter or spirit: this is the dual world we have inhabited ever since—where the brain's ability to distinguish and classify has ruled the roost. From this duality come the ideas we live by, what William Blake called "mind forg'd manacles," the mental abstractions that seem too obvious to question, that construct and confine our vision of reality.

Once upon a time—but this is neither a fairy tale nor a bedtime story— we knew less about the natural world than we do today. Much less. But we understood that world better, much better, for we lived ever so much closer to its rhythms.

Most of us have wandered far from our earlier understanding, from our long-ago intimacy. We take for granted what our ancestors could not, dared not, take for granted; we have set ourselves apart from the world of the seasons, the world of floods and rainbows and new moons. Nor, acknowledging our loss, can we simply reverse course, pretend to innocence in order to rediscover intimacy. Too much has intervened.

—DANIEL SWARTZ, "Jews, Jewish Texts and Nature"

This divided world places each of us as a mind inside the limits of our bodies. This, we believe, is the edge of *me*, this layer of skin; this is the organism I propel through the world, surrounded by things, receiving sensory messages—smells, tastes, sights— through various orifices and nerve endings, which may help me to know the world outside, or may turn out to be dangerous misconceptions. This idea of the body as a machine—quite new in the history of our species—has produced technology to remedy its limits: more machines to extend the reach, accelerate the motion and magnify the strength and sensory acuity of this body-machine as it acts on the world beyond. Mind within body—the ghost in the machine—that is what our culture teaches us we are, what we accept as obvious and normal and real.

> *We milk the cow of the world, and as we do*
> *We whisper in her ear, "You are not true."*
> —RICHARD WILBUR, "Epistemology"

Trapped in this body, we are caught in the very thing we fear most—mortality. Although modern medicine does its best, it too is part of the same cultural worldview; it works within the separated dual world. Modern medicine has no limits or boundaries to guide its path. Instead, its overriding imperative seems to be that if it can be done, it must be done. For medicine, biological limits are a challenge to be overcome—when a 1-kilogram premature infant can be helped to survive, a 0.5-kilogram infant becomes the next possibility. If surgery can correct congenital defects in an infant, then the techniques that are used enable fetal surgery to take place. Caesarean deliveries, hormone-induced multiple ovulation, in vitro fertilization, embryo transplants—they all elevate the scientist-doctor to a directing role in the intimate process of reproduction and development. At the other end of our life, the process of aging is often portrayed as a disease, a breakdown of organ, tissue or genetic systems, a flaw in the machine. Once it is viewed as an abnormality, aging becomes a challenge or a disgrace rather than an honourable and admirable stage in human life.

For perhaps the first time the landscape of meaning is supplanted by the landscape of fact. Before the Renaissance human beings, like other creatures, occupied a qualitatively heterogenous world, riddled with significant places. Only the offspring of the Renaissance have ever imagined it to be all the same, neutral matter for transformation and exploitation. This they accomplished by scraping all traces of value from the environment and vesting it solely within the boundaries of the ego. The result is an aggrandizement of the individual human being and the creation of a bare and bleached environment.

—NEIL EVERNDEN, *The Natural Alien*

Eventually the body must weaken and die—the machine wears out. When it does, the ghost must disappear. These are the consequences not of mortality but of the way we think about it. Divided from each other, we try to make contact beyond our own limited selves, struggle to construct and maintain a community in a world designed around the individual, search for lasting connection. Separated from the natural world, we are lonely, destructive and guilty—but our solutions to environmental destruction are crafted within the frame of mind that created the division and isolation. "Saving nature" because it makes economic sense, because the natural world may contain drugs to heal human ills or even because doing so is "natural justice"— these are all *arguments* from the Cartesian world, where mind acts on the world, observing, analyzing, quantifying. Above all, they are *arguments*, and in every argument there is a winner and a loser.

Cold and austere, proposing no explanation but imposing an ascetic renunciation of all other spiritual fare, this idea [that objective knowledge is the only source of truth] was not of a kind to allay anxiety, but aggravated it instead. By a single stroke it claimed to sweep away the tradition of a hundred thousand years, which had become one with human nature itself. It wrote an end to the ancient animist covenant between man and nature, leaving nothing in the place of that precious bond but an anxious quest in a frozen universe of solitude.

—JACQUES MONOD, *Chance and Necessity*

The scientific method is a refinement of the way we in the Western world learn to see and understand the world from the beginning of life to its end, cleared (as we are taught) of all the confusions and irrelevancies of transitory personal experience. Modern science confirms and re-enacts this picture of reality, examining and exploring nature piece by piece, in the hopes of reassembling it into an intelligible, rational abstract system that contains everything-that-is. It is as if we have embraced the risks of consciousness, sent ourselves into exile to *abstract* the meaning and the value of the world. At the same time we have repudiated the forgotten rituals and sacrifices that would heal the wound.

Today there are many signs that Western culture is starting to reconsider its drive to colonize and exploit the rest of the planet. The search for spiritual reconciliation is taking many different forms: finding virtue in crystals, looking for guidance from the movement of the planets, submitting to the demands of sects and cults, seeking rebirth in new forms of old religions, making pilgrimages, gathering at sacred sites. All these, and many other expeditions into the "supernatural" or "paranormal," represent a widespread, deeply felt longing for wholeness and purpose on this Earth. Theologians, ecologists and Indigenous Peoples are finding common ground as they explore the need to recognize the sacred in the here and now, rather than in the hereafter, and try to help human beings return home to their place in creation.

The knowledge to be gained from Indigenous Peoples worldwide is especially important. Living in place over time offers understanding of how ecosystems work and helps people forge relationships with the land, water, air and all within it. Many Indigenous Peoples refer to everything from trees and rocks to animals as "kin." Deep connections can have a spiritual element, but they're practical as well. We see nature as something we are a part of rather than simply "resources" to exploit. We learn to care for and live within the limits of that which we've come to respect and love. We begin to understand the interconnectedness of all existence.

Returning to the Garden

Only human beings have come to a point where they no longer know why they exist. They... have forgotten the secret knowledge of their bodies, their senses, their dreams.
—LAME DEER, quoted in D.M. Levin, *The Body's Recollection of Being: Phenomenal Psychology and the Deconstruction of Nihilism*

Millions of years of ancestral experience are stored up in the instinctive reactions of organic matter, and in the functions of the body there is incorporated a living knowledge, almost universal in scope...
—ERICH NEUMANN, *The Origins and History of Consciousness*

How can we re-enter the world, restore its spirits and celebrate the sacred? Psychologist David Michael Levin believes we must begin by re-entering the body, exploring the way technology has cut us off from the body's wisdom. All human stories work to weave meaning and order out of disjunction and confusion, but the story told by the Western world specifically excludes human experience as a source of truth. We assert an "objective reality," made of abstract universal principles, which is more correct, more accurate than the messy sensory world we experience daily. But that sensory world is the one we are part of, which penetrates us, and which we create and re-create continually. Just a moment's thought reveals how "subjective" the world actually is. Walk your garden in midsummer and watch how it moves and changes around you. Each plant has a history known to you— where it came from, who brought it to the garden, where else it has grown, how it has thrived. Each plant is dense with relationships made and sustained by your consciousness, like a field of meaning extending through time and space. Flower beds grow other meanings as well: some speak directly to you—of success, perhaps, or reproach for weeding or pruning still undone—others hint at possibilities, at relationships with other parts of the garden world. Do the colours changing on that leaf describe the chemistry of the soil, or do they signal the arrival of another organism—a mildew, perhaps, or a fungus? Ants always crawl on the peony's fat buds; how do they affect the flower? A multitude of living beings, including the gardener, create and maintain that field of meaning—butterflies, birds, beetles, soil organisms, moving and acting in their own sphere, intersecting with each other, intent, purposive, beautiful. Is this "objective reality"? Of course not. Is it reality? Of course.

For if our body is the matter upon which our consciousness applies itself, it is coextensive with our consciousness. It includes everything that we perceive; it extends unto the stars.
—HENRI BERGSON, quoted in D.M. Levin, *The Body's Recollection of Being: Phenomenal Psychology and the Deconstruction of Nihilism*

Attending to our experience, putting spirit back into the fingertips, allows us to redefine consciousness—instead of being trapped inside the mind it becomes a reach, a region of care, the conversation we have with the garden around us. "The conversation of mankind," according to ecologist Joseph Meeker, "is an open and continuing dialogue that connects our bodies and minds intimately with the processes of nature that permeate all life forms." Like any other dialogue, it requires attention, he believes:

Learning to converse well with the world can begin by listening carefully to the messages sent ceaselessly by our bodies and by the other forms of life that share this planet. The best conversations are still those that play variations on that great and ancient theme, "I'm here; Where are you?"

To see a World in a grain of sand,
And a Heaven in a wild flower,
Hold Infinity in the palm of your hand,
And Eternity in an hour.

—WILLIAM BLAKE, "Auguries of Innocence"

We long to escape death, so we reject the mortal body and its communication with the world around it and search for abstract, eternal knowledge. But science—as well as our myths—tells us where immortality lies: it lies in the world we belong to, in the matter we are made from. Matter is not mortal; as we have seen earlier, matter is not transitory, it is transformational, it moves through time and space, from form to form, but it is never lost. We know about this kind of immortality; our intimations of it take a million bodily forms—the curve of the child's head inherited from his great-grandfather, the familiar posture of the woman sweeping, passed down since time immemorial, the hand raised in farewell, the curve of the lips smiling a greeting; these are the genetic and social ways we humans endure forever. But the transformation of our personal matter extends beyond the genetic and social world of humans; the material each one of us is made from comes from and goes to the world around us. In "Transformations" Thomas Hardy expresses a homely version of this endless process:

Portion of this yew
Is a man my grandsire knew,
Bosomed here at its foot:
This branch may be his wife,
A ruddy human life
Now turned to a green shoot...

> So, they are not underground,
> But as nerves and veins abound
> In the growths of upper air,
> And they feel the sun and rain,
> And the energy again
> That made them what they were!

Not dead matter, but alive as nerves and veins, senses and energy, the human couple is still participating in the consciousness, the being of Earth. Re-entering the body reanimates the world around us; the spirits return to the sacred grove.

Obituary, May 8, 1994
Carr Kaoru Suzuki died peacefully on May 8th. He was eighty-five. His ashes will be spread on the winds of Quadra Island. He found great strength in the Japanese tradition of nature-worship. Shortly before he died, he said: "I will return to nature where I came from. I will be part of the fish, the trees, the birds—that's my reincarnation. I have had a rich and full life and have no regrets. I will live on in your memories of me and through my grandchildren."

The Ecological Vision

Today we can see the beginning of a new way of thinking about the world—as sets of relationships rather than separated objects—which we call ecology. We tend to think of a tree as the brown and green bit sticking up above the ground. Even if you include the roots, you are excluding most of the tree. The air that moves around it, the water that moves through it, the sunlight that animates it, the earth that supports it are all integral parts of the tree. What about the insects that fertilize it, the fungi that help it draw in nutrients, and all the rest of the life involved with that tree? Is the visible solidity the only "real" part, or does it exist as process, relationship, connection as well? We know the answer very well; a life-sized model made in materials indistinguishable from those of an actual tree would hardly fool us for a minute— we know a tree when we see one. We know it is more than our cultural definition allows, and today some ecologists are co-opting the language of science to give us a fuller description of our world.

A tree, we might say, is not so much a thing as a rhythm of exchange, or perhaps a centre of organizational forces. Transpiration induces the upward flow of water and dissolved materials, facilitating an inflow from the soil. If we were aware of this rather than the appearance of a tree-form, we might regard the tree as a centre of a force-field to which water is drawn... The object to which we attach significance is the configuration of the forces necessary to being a tree... rigid attention to boundaries can obscure the act of being itself.

   —NEIL EVERNDEN, *The Natural Alien*

This redefinition of something as familiar as a tree at first rings strange. But we can recognize the more-than-tree-form it describes, just as we know that a forest is more than just the trees that grow there, and that our intercourse with the world extends beyond the edges of our skin. Our language falls short of our apprehension because of the way we have been taught to identify the world. We belong to, are made of, that world that surrounds us, and we respond to it in ways beyond knowing.

> *O chestnut-tree, great-rooted blossomer,*
> *Are you the leaf, the blossom or the bole?*
> *O body swayed to music, O brightening glance,*
> *How can we know the dancer from the dance?*
>   —W.B. YEATS, "Among School Children"

The world, you may say, is not a garden, and thinking like a leaf was not the way that humans invented elevators and vaccinations. Indeed, the world we have created is an extraordinary, unprecedented achievement, constructed out of the awesome power of our abstracting, pattern-making brain. But it has lacked the ingredient we discover we depend on to thrive—the idea of wholeness and connection we call spirit. Human beings have always believed in power beyond human power, life after death and spirit-among-us (the sacred, the holy). But our cultural narrative does not include these beliefs, so our experience of them is stunted, truncated, painful. The consequences are threatening indeed—the denial of value, the negation of being. But if we look carefully, we will see that that original story is still telling itself within us and around us, even in our de-spirited culture.

The Spirit Speaks

> Let the heavens rejoice, and let the Earth be glad;
> let the sea roar, and all that fills it;
> let the fields exult, and everything in it!
> Then shall all the trees of the wood sing for joy
> before the LORD, for he comes,
> for he comes to judge the Earth.
> He will judge the world with righteousness,
> and the peoples with his truth.
> —PSALM 96:11–13

The psalm asserts the song of the Earth by singing it. Giving voice to Earth's voice has been a specific human task since the beginning, according to the stories we tell ourselves, the songs we sing, our rituals and our poetry. Repetition, rhythm, rhyme, patterns of gesture, movement and language: these are the ways we speak out and give coherence to experience, assert our connection with everything else. These repetitive, echoing forms of speech and movement shape meaning out of randomness, mimic and embody the cyclic, interdependent processes that create and maintain life on Earth—the web we are part of. In place of the linear time of mortality, dance and poetry beat out a circular measure, keep time with the world.

> Mortals *leap* and *dance* in obedience to the Earth, the elemental presence of ground; mortals leap and dance with a *rhythm* of power, a rhythm which gets its measure from the immeasurable ground which stands under their feet. Being skilful, the gestures of dance are celebratory. They commemorate and give thanks. They surrender the ego's will to power, giving it back, as the acceptance of our mortality, to the all-powerful Earth, ground of our body of understanding.

Human language was the gift of power, an instrument of creation analogous to that of the gods, as Spider Woman suggests, in the Hopi myth at the beginning of this chapter, when she asks for "speech," "wisdom" and "reproduction" for the First People. Similarly, the Creator of Genesis assigns the power of naming to Adam:

> So out of the ground the LORD GOD formed every beast of the field, and every bird of the air, and brought them to the man to see what he would call them; and whatever the man called every living creature, that was its name.

Naming a thing creates an identity; names establish values and functions, give something life, a separate existence. We *are* our names in ways we cannot describe; we hear ourselves called across a noisy room, we feel as though the very letters are

somehow ours. Language weaves worlds of being and meaning; but this is a double-edged sword. Calling a forest "timber,"
fish "resources," the wilderness "raw material" licenses the treatment of them accordingly. The propaganda of destructive forest
practices informs us that "the clearcut is a temporary meadow."
Definition identifies, specifies and limits a thing, describes what
it is and what it is not: it is the tool of our great classifying brain.
Poetry, in contrast, is the tool of synthesis, of narrative. It strug-
gles with boundaries in an effort to mean more, include more, to
find the universal in the particular. It is the dance of words, creating more-than-meaning, reattaching the name, the thing, to
everything around it.

> *My only drink is meaning from the deep brain,*
> *What the birds and the grass and the stones drink.*
> *Let everything flow*
> *Up to the four elements,*
> *Up to water and Earth and fire and air.*
> —SEAMUS HEANEY, "The First Words"

Since poetry began, poets and songwriters have been fight-
ing the mind/body dichotomy, singing their sense of the world, of
the body and spirit moving together through the world eternally.
Poetry takes the fractured, mortal, longing human creature and
reshapes it into be-longing. Crafted words attempt to resolve the
contradictions of consciousness, catching speech (as insubstan-
tial as air, as transitory as breath) as it comes and goes, tying it
into the eternal.

> *And I have felt*
> *A presence that disturbs me with the joy*
> *Of elevated thoughts; a sense sublime*
> *Of something far more deeply interfused,*

Whose dwelling is the light of setting suns,
And the round ocean and the living air,
And the blue sky, and in the mind of man:
A motion and a spirit, that impels
All thinking things, all objects of all thought,
And rolls through all things.

—WILLIAM WORDSWORTH, "Tintern Abbey"

As the Cartesian view of the world tightened its grip on the West, poets, writers and philosophers mounted a counterattack, using their personal experience of nature as a weapon against the abstracting principles of science. In the late eighteenth and early nineteenth centuries in Britain and Europe, the Romantic movement produced extraordinary poetry, then and now often discounted as sentimental or antirational. In fact the best of Romantic poetry was deeply subversive, attacking the conventional wisdom of that time and of ours. Poets such as Blake, Schiller and Wordsworth insisted on the reality and primacy of human perception, and the crucial insight it brings.

Artists of all kinds tend to see their work as similar to, even identical with, the work of the natural world and its continual process of creation. "Great works of art," according to Goethe, "are works of nature just as truly as mountains, streams, and plains." Thomas Huxley, the great Victorian scientist, reversed the comparison: "Living nature is not a mechanism," he insisted, "but a poem." All creation was Paul Klee's source for his paintings: "I sink myself beforehand in the universe and then stand in a brotherly relationship to my neighbours, to everything on this Earth." In many societies art is practical (because it is powerful), a design embedded in the necessities of life: carved posts that hold up the roof and guard the dwelling place, rituals for healing, rain dances, sand paintings. These are ways of rendering visible the designs of the universe, the cables that tie us to Earth. In

our society this intrinsic value has been expunged from art. In
some cases a monetary value has taken its place, as in the bid-
ding wars for Van Gogh's *Irises* or the cost of tickets to the ballet.
In other cases the official story is that the objects or processes are
valueless—the round games of children, the marching songs of
soldiers, bedtime stories, sandcastles (those most contradictory
objects: fortified, embattled structures that will vanish on the
incoming tide). These are just games, we say, just pastimes; but
as we say it, we know quite well another more fundamental truth.
The world is no more, and no less, than a pastime too—the game
of life, matter and spirit playing together. We are players in the
game, voicing it, telling its story. We speak, therefore we communicate. We sing, therefore we join the song of creation.

Beyond Economic Value

Nobel Prizes are awarded to economists who are attempting to put economic value on everything that matters to us, not
just labour and material goods, but human relationships, family,
divorce, children, love and hatred. This phenomenon reinforces
the notion that economics is the dominant feature of modern life
and that nothing lies beyond its reach. Yet there are many things
that are priceless, beyond any reckoning in dollars and cents. Each
of us has such treasures—love letters from the past, a trinket that
belonged to a great-grandparent or favourite aunt, a scrapbook
filled with childhood memories.

A real estate agent's letter urging me to sell my house because
of a heated market made me reflect on what I value about my
house, the things that make it home. The lot is located on the
ocean with a spectacular view of English Bay, downtown Vancouver and the mountains behind West and North Vancouver. But the
things that really matter to me are beyond value. There is a handle

on the gate that my best friend carved when he stayed with us for a week to help me build the fence, and each time I pass through that gate, I think of him. Each year I pick asparagus and raspberries that my father-in-law planted for me because he knew they are my favourites. The English garden he has created is his pride, and every time I pause to enjoy it, I can picture him standing, with his foot on a shovel, puffing his pipe.

We buried the family dog, Pasha, under the dogwood tree in a patch that my daughters have turned into a cemetery for a hamster, a salamander and other dead creatures they've found in the neighbourhood. In the branches of the dogwood is a treehouse I spent many happy hours building and many more watching my children use. A clematis plant has climbed along the back gate. When my mother died, we scattered her ashes on it, and when my sister's daughter died, we added some of her ashes to Grandma's. Now when the purple flowers bloom, the pain of the loss of my mother and niece is softened because I feel they are nearby.

Inside the house we still use a kitchen cabinet that my father made for Tara and me when we were first married. We salvaged it from our apartment, and it is a bit of my father and our early years together. Everywhere throughout the house are reminders of my wife and birthdays, Christmases and Thanksgivings we've shared.

In the real estate market, none of these things adds a cent to the value of the property, yet for me they are what make my house a home and represent value beyond price. What I'm talking about are things that exist only in my mind and heart, memories and experiences that matter to me, that enrich and give meaning to my life. They are spiritual values. No economist will ever be able to factor them into an equation, but they are just as real as and far more important than any amount of money or any material object. □

Living by the Spirit

An ethical system soundly based on ecology is the next crucial step. First we need to reconcile the painful contradictions of our lives. Although we know who we are, where we come from, what we are for, we give that knowledge no weight; our culture tends to deny or conceal that insight, and so we are left alienated and afraid, believing the truth to be "objective" instead of embodied. A world that is raw material, resources, dead matter to be made into things, has nothing sacred in it. So we cut down the sacred grove, lay it waste and declare that it does not matter, because it is only matter. Just so the slavers of an earlier century declared their merchandise to be incapable of "proper human feeling." Just so generations of experimental animals have been sacrificed in the name of research. Pesticides poisoning the lakes and rivers, fish disappearing from the oceans, rainforests going up in smoke— this is the world we have spoken so powerfully into existence, and we will continue to live in it unless we change our tune, tell a different story.

We know very well what matters most to us: the people we love, the place where we live. "Home is where the heart is" embodies the felt truth. We also know what we fear most: separation, loss, exclusion, exile and the final exile, death. Spirituality may be our chiefest local adaptation—the means by which we touch the sacred, hold together against disintegration. The forms and varieties of spiritual belief and ritual among cultures on Earth may be another example of evolution's incredible, extravagant invention of ways for life to survive. We cannot return entirely to that earlier worldview that embedded us so firmly in our ecosystem; too much has intervened. But we might return to some of our oldest questions and find their answers staring us in the face. What is the meaning of life? Answer: life. Why are we here? Answer: to be here, to be-long, to be. The world does many things—it cycles water, builds soil, grows mushrooms, creates bacteria, invents gold, granite, electromagnetic radiation, chestnut trees. And through us it becomes conscious. If we can see (as we once saw very well) that our conversation with the planet is reciprocal and mutually creative, then we cannot help but walk carefully in that field of meaning.

> *If we were not so single-minded*
> *about keeping our lives moving,*
> *and for once could do nothing,*
> *perhaps a huge silence*
> *might interrupt this sadness*
> *of never understanding ourselves*
> *and of threatening ourselves with death.*

> —PABLO NERUDA, quoted in
> R.S. Gottlieb, ed., *This Sacred Earth*

9

Restoring
the Balance

All that we are is story. From the moment we are born to the
time we continue on our spirit journey, we are involved in
the creation of the story of our time here. It is what we arrive with.
It is all we leave behind. We are not the things we accumulate.
We are not the things we deem important. We are story.
All of us. What comes to matter then is the creation of the
best possible story we can while we're here; you, me, us, together.
When we can do that and we take the time to share those stories
with each other, we get bigger inside, we see each other, we
recognize our kinship—we change the world, one story at a time...

—RICHARD WAGAMESE, Anishinaabe author

HUMANITY IS AN infant species, newly evolved from life's
web. And what a magnificent species we are; we can look
out and feel spiritually uplifted by the beauty of a forested valley
or an ice-coated Arctic mountain, we are overwhelmed with awe
at the sight of the star-filled heavens, and we are filled with rev-
erence when we enter a sacred place. In the beauty, mystery and
wonder that our brain perceives and expresses, we add a special
gift to the planet.

But our brash exuberance over our incredible inventiveness and productivity in this century has blinded us to our place

on this planet. If we are to balance and direct our remarkable technological muscle power, we need to regain some ancient virtues: the humility to acknowledge how much we have yet to learn, the respect that will allow us to protect and restore nature, and the love that can lift our eyes to distant horizons, far beyond the next election, paycheque or stock dividend. Above all we need to reclaim our faith in ourselves as creatures of the Earth, living in harmony with all other forms of life.

What a sign of maturity it would be for our species to acknowledge the profound limitations inherent in human knowledge and the destructive consequences of our crude but powerful technologies. It would mark the beginning of wisdom to pay attention to ecosystems delineated by nature—mountain ranges, watersheds, valley bottoms, river and lake systems, wetlands—rather than regions determined by politics or economics. The ebb and flow of organisms—fish, birds, mammals, forests—across the Earth's expanses reflect built-in territorial rhythms that are worthy of respect. The elements that have sparked life onto this planet and continue to fuel it—air, water, soil, energy, biodiversity—are sacrosanct and should be treated as such. There is no ignominy in admitting ignorance or in confessing our inability to manage wild things, to control the forces of nature or even to grasp the cosmic forces that shape our lives. Recognizing and accepting these limitations with humility is the birth of wisdom and the beginning of hope that we will finally rediscover our place in the natural order.

When we acknowledge our dependence on the same biophysical factors that support all other life forms, believing that we have the responsibility for "managing" all of it becomes a terrible burden. But if we look at the world through the lenses of all of life together, we may recognize the origins of our destructive path and realize that we are not the "managers"; there is wisdom enough for self-management in the web of living creatures that has survived for more than 3.6 billion years. Instead of trying and failing to manage the life-support systems of the planet, we—each one of us—can manage the effect we have on those systems.

Knowing how to act is the first big problem. Many people who are eager to work toward personal and public change feel increasingly baffled by the often contradictory messages from experts, as well as the mantras repeated over and over by the media. We no longer trust our innate common sense or the wisdom of our elders. When I asked the great Nisga'a leader James Gosnell how he felt the first time he saw a clearcut, he replied: "I couldn't breathe. It was as if the Earth had been skinned. I couldn't believe anyone would do that to the Earth." Gosnell knew, as most of us do, in "the pit of his stomach," that a vast clearcut where there was once an ancient forest is an affront to life itself. But while Gosnell trusted his own good sense that such practices are a sacrilege, we feel reassured by forestry "experts" who say that "it only looks bad for a little while, but it will soon 'green up.'" These experts call tree plantations "forests," assure us that the animals will flourish in them, tell us the trees can be "harvested" in less than a century and confidently state that this is "proper silvicultural practice." But deep inside, we know otherwise. And what we know is the truth.

When we see a creek that has been straightened and its banks entombed in concrete, when we use a product once and throw it away, when we consume a new molecule that mimics fat but passes through our bodies undigested, when we hear about cloning a sheep or a human being or when we learn about pigs being engineered to provide organs to transplant in humans, we know viscerally that this is not right. When we express such misgivings, however, we are often told that we are "too emotional," as if caring enough to be emotional about an issue somehow invalidates our concern. Or we are told we lack the expertise to make a judgment. We should be confident in our initial gut reactions and insist that the experts prove their case.

What Can We Do?

At this critical juncture in our history on Earth, we are asking the wrong questions. Instead of "How do we reduce the deficit?" or "How do we carve out a niche in the global economy?" we should be asking, "What is an economy for?" and "How much is enough?" What are the things in life that provide joy and happiness, peace of mind and satisfaction? Does the plethora of goods that our high-production economy delivers so effectively provide the route to happiness and satisfaction, or do the relationships between human and non-human beings still form the core of the important things in life? Is the uniformity of food and other products that we now encounter everywhere on the globe an adequate substitute for the different and the unexpected? We seem to have forgotten the real things that matter and must establish the real bottom line of non-negotiable needs in order to regain a balance with our surroundings. To that end, there are many things that each of us can do. Here are just a few, very practical approaches to changing the way you think and live.

- Think critically about the information that floods over us. Consider its source carefully. Organizations that are funded by vested interests such as the petrochemical industry, forest companies or the tobacco business may not be reliable. Environmental activists have had an effect on corporations, many of which have adopted ways to decrease pollution or energy use. Even energy giants like BP have acknowledged the hazards of climate change and have committed to substantially reducing emissions from their operations and products. But some companies have launched slick public relations campaigns to imply they are environmentally responsible, thereby "greenwashing" themselves while doing little to reduce their environmental impacts. Often a spokesperson for such a group may claim credibility because of a background in environmental work. But after betraying Jesus, Judas would have had little credibility arguing, "Trust me. I was one of the original disciples." The reason non-governmental organizations, or grassroots groups, have such credibility is that their motives are obvious. They are not after maximum profit, market share or power; they are working for sustainable communities, a future for their children, a clean environment or the protection of wilderness.

· Trust your common sense, your ability to assess information. There is a difference between information published in a tabloid such as the *National Enquirer* and information in magazines such as *Scientific American* or the *New Scientist*. Determining the accuracy and biases of the ever-growing range of online sources has become especially important. Reports in newspapers such as the *Wall Street Journal* (U.S.), the *Globe and Mail* (Canada) or the *Australian* must be read with an understanding of their known pro-business biases. Many books have a strong anti-environmental bias but proclaim their objectivity and balance. For an excellent rebuttal and exposé of many of these books, read Paul and Anne Ehrlich's *Betrayal of Science and Reason*.

· Confirm to your own satisfaction the depths of the global ecological crisis alluded to in the 1992 "World Scientists' Warning to Humanity" and the updated 2017 "World Scientists' Warning to Humanity: A Second Notice," as well as the 2019 "World Scientists' Warning of a Climate Emergency" and the Intergovernmental Panel on Climate Change reports. You needn't be confused about which "expert" to believe; just talk to elders around you, people who have lived in your part of the world for the past seventy or eighty years. Ask them what they remember about the air, about other species, about the water, about neighbourhoods and communities, about caring between people and the ways they communicated and entertained each other. Our elders tell us of the immense changes that have occurred in the span of a single human life; all you have to do is to project the rate of change they have experienced into the future to get an idea of what might be left in the coming decades. Is this progress? Is this way of life sustainable?

• Project your mind far ahead into the future and consider the problems that we are leaving as a legacy for our children and grandchildren. What will the quality of their air, water and soil be like? What kind of food will they eat? How much wilderness will be left for them to enjoy? If we can't act with seriousness to alleviate problems of toxic pollution, deforestation or climate change, will future generations be able to overcome the consequences we leave for them?

• Think deeply about some of our most widely held assumptions; many underlie the destructive path we're on.
   – It is widely believed that human beings are special, that our intellect has lifted us out of the natural world into a human-created environment. Our absolute need for air, water, soil, energy and biodiversity belies that assumption.
   – These days, many people believe that science and technology provide us with the understanding and tools to manage nature and to find solutions to problems that science and technology have helped to create. Technology provides powerful but crude instruments for very straightforward activities. But science fragments the way we see the world, so we

have no context within which to see what impact our activities and technological applications have.

- Some people believe that a clean environment is only affordable when the economy is strong, but in fact, it's the other way around; the biosphere is what gives us life and a living. Human beings and our economies have to find a place within the environment. The economic assumption that endless growth is not only necessary but possible is suicidal for any species that lives in a finite world.

- We assume that even though we are just one of perhaps 30 million species, the entire planet is ours for the taking. We assume that we can manage our natural resources through the bureaucratic subdivisions of government and industry. We assume that we can do environmental assessments and cost-benefit analyses to minimize the impact of what we do. All of these assumptions, and many more, fail to stand up to critical analysis yet are seldom challenged or questioned.

· Think about your connections with the living web. How does your use of a car to drive two blocks to the store affect the planet? What are the true costs and benefits of having an article of clothing that is fashionable for the moment and soon discarded?

· Construct a hierarchy of your own basic needs, those things that are absolutely necessary for you to live, to be fulfilled and to be happy. You can do that by answering some pretty basic questions. Is spending time with people you love more important than going shopping? Is making more money, owning a bigger car or possessing the latest technological gadget necessary to fulfill your real biological, social and spiritual needs? What is the value of community, justice, wilderness and species diversity to you, your children and their grandchildren?

This book has attempted to show that we have a hierarchy of at least three tiers of non-negotiable needs. The first is the group of factors that meet our biological requirements—clean air, clean water, clean soil and food, energy and biodiversity. Unless these fundamental requirements are fulfilled, people will not be able to look beyond satisfying them immediately at any cost. A second tier of requirements is the needs coming from our social nature. In order to lead full, rich lives, we need love above all else, and the best way to satisfy that need is to provide for the stability of families and communities. In order to develop one's fullest potential, we must be assured of meaningful employment, justice and security, for without them, we become crippled and incomplete. And finally, as spiritual beings, we have to know that there are forces in the cosmos beyond human understanding and control, that we are indissolubly part of the totality of life on Earth, caught up in an endless process of creation. Only by meeting all of these levels of needs can society provide full satisfaction and opportunity to its members and achieve true sustainability.

· Reflect on how we can meet our fundamental needs while also making a living. To do so means that our economy has to be connected to the real world of the biosphere. Globalization means that in Vancouver, lamb grown in and shipped from New Zealand is cheaper than sheep raised in the Fraser Valley 100 kilometres away. This may make economic sense, but it does not make ecological sense. Estimates of the annual cost of "services" performed by nature, such as cleansing air and water, pollinating plants, inhibiting erosion and flooding, building topsoil and so on, come to the tens of trillions of dollars, yet conventional economics considers these gifts from nature "externalities," meaning they are not included in the economic calculations. A true Earth economy would take these services into account. The notion that growth is the definition of progress is suicidal. As ecologist Paul Ehrlich says, endless growth in a finite world is the creed of the cancer cell and the result of adhering to it is death.

· Our attempts to ward off the negative consequences of our activity must be made at "the top of the pipe," not at the end. That is, we must try to avoid problems such as pollution or climate change by preventing the cause of the problem in the first place rather than by trying to solve the problem after it exists. Our system of costing human products and activities must include the ecological costs as well as costs from extraction all the way through to the final discarding—so-called cradle-to-grave cost accounting; current economic practice omits many relevant costs.

· Protect the vigour and diversity of local communities. The social unit that will have the greatest stability and resilience into the future is the local community, which provides individuals and families with a sense of place and belonging, fellowship and support, purpose and meaning. The local community provides a common history and culture, shared values and a shared future. It is not surprising that when some Indigenous communities have been presented with a "ten-year logging plan" from a forest company, they have rejected it with the demand that the company "come back with a five-hundred-year logging plan." That kind of perspective comes from a community that is profoundly rooted in and committed to a place.

How do we provide the kinds of community services we expect while building on the foundation of our basic needs? A community that wishes to avoid the cycle of boom and bust that results from short-term exploitation of a resource or industrial opportunity must provide long-term, steady employment options, relative freedom from the uncertainty of the global economy and market forces, and most vital services from its own resources and personnel. It is important to support the local community to keep it vibrant, because without strong local communities we certainly will not be able to save the mosaic of all communities on the planet. Some simple homilies for how to support the local community are shop locally, eat locally and seasonally, hire locally, work locally, play locally.

- We can begin to reconnect ourselves to everything on Earth, re-creating a complete worldview by establishing or rediscovering rituals and ceremonies that celebrate those linkages and our communities. We have many rituals that could be built upon—Thanksgiving, Halloween, seasonal festivals celebrating fruits, vegetables, water or some special local feature.

- We can adopt more holistic ways of seeing by learning about and supporting Indigenous history, knowledge, rights, reconciliation and movements. In challenging and overcoming ingrained colonialist perspectives, we can share in the necessary wisdom and knowledge accumulated by people in place over millennia—and we can learn about the historical injustices that for too long have tried to deny that knowledge and wisdom. Acknowledging Indigenous territories has become common practice before meetings and events in Canada, Australia and New Zealand. It's a start, but we all need to do more to create more equitable and enlightened societies based on respect for and responsibility to each other and all existence.

- Get involved. In the long run, sustainable living demands a fundamental shift in values. But action invariably precedes a profound shift in values, so actually doing something is important. In the process, one learns and becomes committed.

    At the most superficial but nevertheless important level, support an environmental organization with money. Most operate on shoestring budgets, and even small donations go a long way. Check out the environmental groups around you and find one (or several) that reflects your values and support it (or them). Volunteer your services, not just to an environmental organization but to one of the many good groups that are concerned with meeting the needs of family and community. Most non-governmental organizations depend on volunteers, and it is both enjoyable and educational to meet and work with them. I have been astounded at the number and quality of volunteers at my own foundation. I have been a full-time volunteer there ever since it was started, and it has been one of the most satisfying and rewarding activities of my life. Volunteering gives you a sense of actually working for the benefit of humankind and the future. And it rewards you with a great deal of fun, especially when you find companions who share your goals and are working along with you to achieve them.

I encounter people who say, "How do you know you're having any effect?" or who sigh with resignation, "What's the point? We're insignificant." Giving up is a cop-out because no one knows what the future holds. Among almost 8 billion people, what each of us does seems insignificant, but a lot of insignificant people can add up to a real force. The David Suzuki Foundation has long worked to determine the most effective steps that we as individuals can take to reduce our negative impact on our surroundings. We focused on housing, transportation and food, the three areas where our choices and actions have the greatest repercussions. The ten suggestions are:

1. Urge government to take bold, ambitious climate action now.
2. Use energy wisely—and save money too!
3. Get charged up with renewables.
4. Eat for a climate-stable planet.

5. Start a climate conversation.
6. Green your commute.
7. Consume less, waste less, enjoy life more.
8. Invest in renewables and divest from fossil fuels.
9. Mobilize for local climate action.
10. Get politically active and vote.

For me, the rewards of getting involved are that I will be able to look my children in the eye and say, "I did the best I could." You see, I believe what my father taught me—that I am what I *do*, not what I *say*. And if everyone works hard and we are lucky, I might even be able to reassure my children, as I used to when they woke up after a bad dream, that "everything is going to be all right." Right now we cannot say that with much confidence.

• Work to get your home as ecologically benign as possible. Of the three Rs—reduce, reuse, recycle—reducing is by far the most important precept. Begin by eliminating the notion that throw-away items are acceptable. Make "disposable" an obscene word and favour the reusable over the recyclable. So much of our economy is based on the consumption of what are essentially trinkets, useless items that titillate briefly and then are forgotten or thrown away. The proliferation of consumer goods is driven by the need to find a niche in the market rather than to satisfy a genuine human need, so, as consumers, we can register a protest by not buying in. I hasten to add that a sustainable world does not have to be drab and monotonous, devoid of colour or frivolity. It is one in which products are made to service a real human need and incorporate principles of durability, recyclability and low energy and ecological cost. Those are principles that must be taken for granted before designing products that are fun and attractive.

Living in a way that satisfies our biological, social and spiritual needs does not have to be filled with pain and sacrifice, a cheerless life without fun, frivolity or laughter. Indeed, I believe time spent with those we love, sharing conversation, companionship and activities, is far more pleasurable than the fleeting enjoyment of possessing a novelty item or the sensory overload of virtual reality.

· There are many small ways to modify your lifestyle that are good for your health, your pocketbook and the health of the planet. They are just common-sense notions that were tacitly assumed two generations ago. Acting on them reinforces our understanding that we can live in a way that makes Earth sense.
  – Ask, "Do I really need this?" before purchasing an item.
  – Take public transit to work whenever possible.
  – Walk, bike or rollerblade if you have to make a trip of less than ten blocks.
  – Use both sides of sheets of paper.
  – Make garbage-free lunches for your children.
  – There are many books with helpful tips on ways to live more lightly on the planet. Buy one and begin to do what it suggests.

· Industries that are designing means of production can follow the example of nature in which one species' waste is another's opportunity. A plant that utilizes energy from the sun to grow and reproduce may also nourish a host of parasites and herbivores and upon dying feed still other life forms while returning organic material to the soil to nurture future generations of plants. Material is used, transformed and used again in a never-ending cycle. Nature has evolved exquisite mechanisms to survive and flourish over the vast expanse of time as the Earth has undergone geological and climatological change. Our scientific insights are profound, but we are still at an early stage of unlocking nature's secrets when technology is powerful but crude. Too often we lack the knowledge to anticipate the consequences of our manipulations, as we have learned with nuclear energy, DDT, CFCs and GMOs. But we need science and technology to reveal the nature of the challenges that confront us and minimize the dangers ahead. We need far more basic research and the application of technology with a sense of humility and respect for nature. Janine Benyus has suggested that we use what she calls biomimicry—copying nature rather than bludgeoning our surroundings into submission—as a principle to guide our actions. We can transform our thinking from the linearity of extracting, processing, manufacturing, selling, using and discarding into the circularity of natural cycles.

- Go out into nature. Nature is not our enemy; it is our home. In fact, it sustains us and is in every one of us. All living things are our relatives and belong with us in the biosphere. Out of doors we learn very quickly that there is another rhythm and a different agenda from the frenetic human pace and program. Feel the rain and wind on your face, smell the fragrance of soil and ocean, gaze at the spectacle of the myriad stars in clear air or countless animals making their annual migration. Doing so will rekindle that sense of wonder and excitement we all had as children discovering the world and will engender a feeling of peace and harmony at being in balance with the natural world that is our home.

- Don't feel guilty. Guilt is draining and oppressive—as the T-shirt says, Nobody's Percfet. I have radically reduced my use of the car, but I have taken many airplane flights, and each flight releases huge amounts of pollutants and greenhouse gases on my behalf. Although I have given up red meat, I eat fish. There are many things that I live with that prevent me from being environmentally pure. The most important thing now is the exchange of ideas, to spread the word as we all work toward reducing our effect on the planet, developing infrastructures so that we can live sustainably and creating public support that will ultimately change political priorities. That's the important work right now. But there are things that we can do while we make the transition to sustainable living. For example, I purchased the first gas/electric hybrid car sold in Canada, the Prius, and now own a fully electric vehicle. I also try to avoid travelling by plane and purchase "carbon credits" when I do fly or travel by car. These put clean energy such as wind or solar energy onto the electrical grid to balance the emissions made on my behalf, but I must still reduce my carbon emissions.

Until one is committed there is hesitancy, the chance to draw back, always ineffectiveness, concerning all acts of initiation (and creation)... the moment one definitely commits, then Providence comes too. All sorts of things occur to help one that would never otherwise have occurred... Whatever you can do or dream you can, begin it. Boldness has genius, power and magic in it. Begin it now.

—GOETHE, quoted in P. Crean and P. Kome, eds.,
Peace, a Dream Unfolding

Adding Up to an Irresistible Force

Lest anyone despair, it is worth remembering Margaret Mead's words: "Never doubt that a small group of thoughtful, committed citizens can change the world."

In 1985, anyone who speculated aloud that the Soviet Union would soon cease to exist and bring an end to the arms race, that the Berlin Wall would come down so that two Germanys would become one, and that apartheid would end and Nelson Mandela would get out of jail and become president of South Africa would have been certified as crazy. Yet less than ten years later, the inconceivable happened with very little bloodshed. One cannot give up hope; incredible things do happen, and no one can know what the final contributing factor might be. Consider Al Gore, former vice-president of the U.S. and failed presidential candidate. He had signalled his interest in and concern about the environment with his bestselling book *Earth in the Balance*, and after a shattering political loss, despite garnering a majority of votes, Gore continued speaking out on climate change. With the release of the film *An Inconvenient Truth*, and the book based on it, Gore achieved global recognition and respect for his call to action to combat the threats of global warming. In *Earth in the Balance*, Al Gore described the results of research on the physical properties of growing sand piles. When grains of sand are added to a pile one at a time, the pile grows until it reaches a critical point at which the addition of one more grain of sand causes avalanches, slides and massive changes. It is an apt metaphor for the way individuals can create sudden shifts in popular understanding and social action. No one can predict when that critical point will be reached when one additional grain can be the final agent that will cause an enormous shift. Each person, group or organization working toward a different world may seem powerless and insignificant, but all of them can add up to a force that can become irresistible.

In 1972, when the United Nations convened the first global conference on the environment, only about ten national government institutions in the world dealt with the environment. Today almost every government at federal, state and municipal levels has environment departments or committees. Virtually all of the delegates at the meeting were from industrialized countries. Today, throughout the developing world, there are many grassroots and non-governmental environmental groups. Indonesia alone has more than a thousand, and from Brazil to Malaysia,

Japan and Kenya, there are grassroots organizations and groups working toward a sustainable way of living.

Today any magazine or journal with an environmental focus features good-news stories. And there are hundreds, if not thousands, of such reports. Environmental awards such as the annual Goldman Environmental Prize, the Champions of the Earth awards, the Blue Planet Prize and so on honour a few of hundreds of deserving nominees. This is the basis for real hope that change is happening.

Environmentalists are often accused of being "negative," "against development," "anti-progress" or "too depressing." So at the turn of the millennium, researcher Holly Dressel and I set out to determine whether there are alternatives to practices that are destructive to the environment. We looked for what individuals, companies, organizations and governments are doing to genuinely seek alternative, sustainable practices. We anticipated a slender book. To our surprise and delight, we could have written volumes. We called the book *Good News for a Change: Hope for a Troubled Planet,* and it documented many activities in countries both rich and poor that seek to find a balance with nature. The thousands of positive stories to report from all over the world occur at the level of nations, states and provinces, municipalities, companies and groups. But most profoundly, individuals are being impelled to act; in doing so they have an effect not only locally but often right across the globe. What follows is a completely personal selection of just a few of the many stories that have inspired me. They show just how much power each one of us has over the Earth's future.

A Child's Reminder at the Earth Summit

... and a little child shall lead them.
—ISAIAH 11:6

I'm only a child, yet I know if all the money spent on war was spent on ending poverty and finding environmental answers, what a wonderful place this Earth would be...
—SEVERN CULLIS-SUZUKI, age twelve, Earth Summit,
Rio de Janeiro, June 1992

My family's ten-day visit to the village of Aucre in the Brazilian rainforest took place in 1989, when my daughters Severn and Sarika were nine and five years old, respectively. As we flew away from the village, we could see the encroachment of gold miners who were polluting the rivers and destroying the riverbanks, and the farmers burning the forest down in a desperate search for land to grow food on. Severn became alarmed about the future of her new-found friends in Aucre, and upon returning to Vancouver, she started a club called ECO (Environmental Children's Organization). Five ten-year-old girls began to speak out about the beauty of tropical forests; the animals, plants and people who inhabit them; and the need to protect them. Over time, they were invited to visit classes and give talks, gaining some local notoriety.

In 1991, Severn told me that she wanted to take ECO to the Earth Summit in Rio de Janeiro in June 1992. This conference would bring together the largest gathering of heads of state in history. "I think all those grown-ups will be talking about our future," Severn said, "and they need us there to act as their conscience." Two months later Severn proudly displayed a cheque made out to ECO for a thousand dollars from an American philanthropist to whom she had spoken about her dream.

My wife, Tara, and I then realized that the perspective of children was critical for people meeting at the Earth Summit to hear. So we offered to match every dollar the club raised with another. Undaunted by the challenge, the girls set about making and selling ornamental salamanders (eco-geckos) from plastic, selling used books and organizing bake sales. They attracted the attention of another philanthropist, and of Raffi, the well-known children's singer. Both donated generously. Finally, the girls held a public event to show slides and describe their goal. To our amazement, ECO raised over $13,000—enough, with our matching funds, to take five children and three adults to Rio.

At the Earth Summit ECO registered as a non-governmental organization and rented a booth at the Global Forum along with hundreds of other groups. They set up a display of pictures and posters, handed out newspapers and brochures about ECO and talked to many people. Soon reporters and television cameras appeared to interview these five girls from Canada. The Canadian environment minister, Jean Charest, made an appearance with cameras in tow. Eventually James Grant, the U.S. head of UNICEF, heard the girls speak and urged Maurice Strong, the organizer of the Earth Summit, to invite Severn to address a plenary session, and he did.

Severn was twelve years old. She wrote her speech, with input from her fellow ECO members, and rehearsed it over and over in the taxi to Rio Centro, the site of the conference. She was the last to speak, in a huge room that swallowed the few hundred delegates. Some of what she said was this:

> I'm only a child and I don't have all the solutions, but I want you
> to realize, neither do you. You don't know how to fix the holes in
> the ozone layer. You don't know how to bring the salmon back
> up a dead stream. You don't know how to bring back an animal
> now extinct. And you can't bring back a forest where there is now
> a desert. If you don't know how to fix it, please stop breaking it...
>
> In my country we make so much waste; we buy and throw
> away, buy and throw away. Yet northern countries will not
> share with the needy. Even when we have more than enough, we
> are afraid to lose some of our wealth, afraid to let go...

You teach us how to behave in the world. You teach us not to fight with others; to work things out; to respect others; to clean up our mess; not to hurt other creatures; to share, not be greedy. Then why do you go out and do the things you tell us not to do?...

My dad always says, "You are what you *do*, not what you *say*." Well, what you do makes me cry at night. You grown-ups say you love us. I challenge you. Please, make your actions reflect your words.

Spoken from the heart, these words electrified the conference room and quickly spread beyond. When the Earth Summit ended, Maurice Strong quoted Severn to remind the delegates why they were there.

Severn is now executive director of the organization her mom and I founded, the David Suzuki Foundation, and a new generation of young people is taking up the fight. I was thrilled in late 2019 when young climate activist Greta Thunberg visited Canada and marched with Severn at the Vancouver climate march. In September, before coming to Vancouver, Greta addressed close to half a million young people in Montreal at the largest march in Quebec's history. At the Vancouver strike, fifteen young people also introduced the legal case they brought against Canada's government for not doing enough to address climate change.

Greta's involvement with climate issues began in 2018, when she held a solitary "school strike for climate" in front of Sweden's parliament. She was fifteen. Within months, young people from around the world were holding their own climate strikes, and one year after her first solitary strike, millions participated in 2500 events in 163 countries on all seven continents.

In a powerful speech at the 2019 World Economic Forum in Davos, Switzerland, Greta told delegates, "Adults keep saying: 'We owe it to the young people to give them hope.' But I don't want your hope. I don't want you to be hopeful. I want you to panic. I want you to feel the fear I feel every day. And then I want you to act. I want you to act as you would in a crisis. I want you to act as if our house is on fire. Because it is."

Children shouldn't have to spend their time in court or protesting. They should be enjoying their formative years, getting outside, playing, spending time with friends and family, studying—even connecting on social media. But we have failed them, and they are doing everything they can to overcome that failure. Because most are too young to vote, they have few avenues other than protests and legal action to get politicians and others to pay attention. Beyond the climate strikes, youth from countries including Canada, the U.S., the U.K., the Netherlands, Portugal, Ireland, Germany and more have launched court challenges against their governments for failing to address the climate emergency.

In Canada, fifteen young people from seven provinces and one territory filed suit, claiming the federal government is contributing to dangerous climate change and in doing so is violating their rights to life, liberty and security of the person under section 7 of the Canadian Charter of Rights and Freedoms and for failing to protect essential public trust resources. The youth plaintiffs also alleged that Canada's conduct violates their right to equality under section 15 of the Charter, since youth are disproportionately affected by the effects of climate change.

These young people are asking only that we heed the scientific evidence and act on it. We should have listened to Severn back in 1992. We should be listening to Greta and all the youth now. Every day we fail to act on the climate crisis is a day stolen from young people and those not yet born. We owe Greta and all young people a debt of gratitude for holding up a mirror to our actions. More than anything, we owe them a future, and that means getting serious about changes needed to resolve the climate crisis.

Ian Kiernan and His Fateful Yacht Race

When I sailed around the world, I saw with my own eyes that people were using the oceans as a trashbin. By the time I got home, I had made up my mind to do something about it, starting with my own backyard, Sydney Harbour.

—IAN KIERNAN, interview with David Suzuki

Ian Kiernan was a whirlwind of activity from Sydney, Australia. He was a professional yachtsman whose life was transformed in 1986–87, when he joined twenty-four people from eleven countries to compete in a round-the-world solo yacht race. Kiernan's 42,000-kilometre trip on a 20-metre boat began in Newport, Rhode Island. It took him across the Atlantic Ocean to Cape Town, South Africa, and then on around Antarctica to Sydney.

In earlier races, Kiernan had always dumped his garbage overboard, but this one was being followed by 33,000 American schoolchildren, so the participants agreed to stow their garbage on board. It shamed Kiernan to realize what he had done in the past; as he spent long hours watching the sea go by, he saw that even in the most remote, pristine waters, human-created debris, particularly Styrofoam and plastic, was everywhere. In his youth, Kiernan had heard about the vast expanse of seaweed filled with life in the Sargasso Sea, and he had looked forward to seeing it first-hand during this race. To his dismay, the first things he encountered were a rubber sandal, a plastic bag and a plastic pipe. In his disappointment and anger, he decided to do something about the mess, beginning in his backyard—Sydney Harbour.

By the time Kiernan reached Sydney, he was a national hero to Australians, and he used his fame to spread his idea of cleaning up the harbour. He put 85 per cent of the money he raised into communicating his idea. In March 1989, he held his first cleanup of Sydney Harbour, hoping a few thousand people might show up to help. To his astonishment, forty thousand people came and collected 5000 tonnes of rubbish! Fortuitously, on the day of the cleanup, the press revealed that raw sewage was being dumped into Sydney Harbour, and the news fuelled the urge of individual citizens to take immediate action.

This sensational event galvanized Australia, and calls began to pour in from other cities. Kiernan went to Tasmania to offer help, to Darwin in the Northern Territory, to Wollongong, where people pulled 158 cars and 2 buses from Lake Illawarra. This response sent a message to politicians that people cared passionately about their environment. Businesses saw opportunities to be good corporate citizens and make money by getting involved. In 1990, Kiernan expanded his project to the entire country, and Clean Up Australia Day attracted 300,000 people. Now every first Sunday in March is Clean Up Australia Day, involving more than half a million people and 850 cities and towns. Kiernan's idea was expanded in 1993 to Clean Up the World, and the United Nations Environment Programme helps show people how to get their own Clean-Up committees going. By 1996, 110 countries were involved and 40 million people took part around the world. Kiernan noted that people in affluent countries seem to care less than those in poorer nations, where cleaning up is a matter of children's health. In Africa, Korea, Poland and Russia, Clean Up programs are under way. Although he has since passed away, from his lonely vigil in the planet's oceans, Kiernan turned his shame and anger into an action plan that has allowed people to take responsibility for their local environment.

A New Kind of Architect: William McDonough

... design leads to the manifestation of human intention, and if what we make with our hands is to be sacred and honor the Earth that gives us life, then the things we make must not only rise from the ground but

return to it—soil to soil, water to water—so everything that is received from the Earth can be freely given back without causing harm to any living system. This is ecology. This is good design.

— WILLIAM MCDONOUGH, quoted in K. Ausubel,

Restoring the Earth

In nature, some animals gather materials from their surroundings to use for survival just as we do. The caddis fly larva incorporates sand or twigs into a protective case; the decorator crab deliberately places anemones or seaweed on its carapace to camouflage itself; gorillas carefully construct sleeping platforms with branches, vines and leaves. But human beings differ from all other species in the way we extract material from our surroundings and exploit it for clothing, shelter and goods and in the amount we extract.

We admire human-created objects that combine form and function in an aesthetically pleasing way—the elegant curve of a samurai sword, the Concorde supersonic jet, the Eiffel Tower or St. Paul's Cathedral. The integration of appearance and efficiency in design is intuitively satisfying. But modern architecture, according to William McDonough, an architect based in Charlottesville, Virginia, blossomed in an era of cheap energy and cheap materials such as glass. As a consequence, our buildings are impressive in appearance but wasteful of energy and filled with toxic chemicals venting from glues, carpets and furniture. McDonough says: "Our culture has adopted a design strategem that says, 'If brute force doesn't work, you're not using enough of it.'"

McDonough is a new kind of architect. His attitude and values were shaped by his upbringing in Hong Kong, where chronic water shortages were a constant reminder of finite resources on Earth. He graduated with a degree in architecture from Yale University in 1976. It was just two years after the Arab oil embargo had shocked the world, yet the best designers seemed to be ignoring the fact of shortages. Critical of the short-sighted view of his profession, McDonough likes to cite a story related by the eco-philosopher Gregory Bateson. At New College in Oxford, England, the huge oak beams of the university's main hall are some 12 metres long and 0.5 metres thick. In 1985, dry rot had finally weakened them so much that they needed to be replaced. If oak trees of such size could have been found in England, they would have cost about US$250,000 per log for a total replacement cost of around US$50 million. Then the university forester informed the administrators that when the main hall had been built 350 years earlier, the architects had instructed that a grove of oak trees be planted and maintained so that when dry rot set in, about three and a half centuries later, the beams could be replaced. Now that is long-term planning, and McDonough believes this has to become standard in architectural thinking.

McDonough has added another R to the trio of reduce, reuse and recycle—*redesign*. In order to avoid the toxicity of today's building materials, he searches tirelessly for non-toxic products. He points out that 30 per cent of material in landfills is construction waste and that 54 per cent of energy in the United States is expended on behalf of construction. Architectural design based on principles of sustainability could have a huge effect on both waste disposal and energy use.

In 1986 McDonough was hired by the Environmental Defense Fund to design a building that would combine conservation and design. It set new levels in energy efficiency, increased the flow of air six times above the normal standard, replaced synthetic materials with natural fibres such as jute and substituted nails for glues. A major concern for McDonough is the plethora of building materials and supplies that vent toxic gases. "We're dripping poison," he says. "We are making products or subcomponents of products that no one should buy." He has been instrumental in searching for ecologically benign materials. The criteria on which he bases acceptability are: "... no mutagens, no carcinogens, no persistent toxins, no heavy metals, no bioaccumulatives, no endocrine disruptors. Period." Out of more than eight thousand chemicals screened, just thirty-eight met the criteria and have the potential of being made into products that are so benign that not only can they be composted, they are also edible.

McDonough got a big chance to spread his ideas when he was asked to design a new Walmart building in Tulsa, Oklahoma. Megastores such as Walmart are often opposed on ecological and social grounds by small merchants and people concerned about local communities, but McDonough embraced the opportunity. Walmart builds a new store somewhere every two days, and that involves a huge amount of building materials and products—for example, almost a kilometre of skylights a day. To McDonough, this meant that if ecological principles were incorporated into Walmart's building program, the very scale of the company's needs could be enough to create new manufacturers of eco-products. Walmart stores are built on the assumption that they will last as stores for forty years, so McDonough designed the Tulsa store to be made over into an apartment building after that period. No CFCs were used in the store, and the introduction of skylights reduced energy needs by 54 per cent. Instead of steel, which requires a lot of energy to create, he used wood in the roof. To ensure that wood came from sustainable forestry practices, he began Woods of the World, a non-profit forestry research information system whose database was made available to architects and builders.

McDonough believes that ecological thinking and sustainability must become as natural a part of life as breathing:

Everybody has to be a designer. What we're calling for is massive creativity. The fundamental issue we're trying to address is the rightful place of humans in the natural world. How can we go about being part of the natural design?

McDonough believes that all sustainability is based in local communities and rests on the recognition that everything is interdependent. Charlottesville is the home of Thomas Jefferson, framer of the American Constitution and the Declaration of Independence, and McDonough plans to honour him:

> ... we're going to declare *interdependence*. Jefferson would do it if he were alive today. It's a declaration for life, liberty and the pursuit of happiness free from remote tyranny. Except this time it's intergenerational remote tyranny, and the tyrants are us and our bad design.

In 2002, McDonough and German chemist Michael Braungart published *Cradle to Cradle: Remaking the Way We Make Things*, which attempts to put us into the next industrial revolution—the deliberate and intelligent design of technology that is sustainable.

I believe we can accomplish great and profitable things within a new conceptual framework—one that values our legacy, honors diversity, and feeds ecosystems and societies... It is time for designs that are creative, abundant, prosperous, and intelligent from the start.

—WILLIAM MCDONOUGH

Reforesting Kenya: Wangari Maathai

All of us have a God in us, and that God is the spirit that unites all life, everything that is on this planet. It must be this voice that is telling me to do something, and I am sure it's the same voice that is speaking to everybody on this planet—at least everybody who seems to be concerned about the fate of the world, the fate of this planet.

—WANGARI MAATHAI, quoted in P. Sears, "Wangari Maathai: 'You Strike the Woman...'"

Wangari Maathai, who died in 2011, stood out in a crowd, especially in North America, for she was statuesque, dressed in bright African patterns and had skin that was almost jet black. She was a remarkable woman by any society's standards. She earned a BA from the United States, an MA from Germany and a PhD from Kenya, and was divorced by her husband for being "too educated, too strong, too successful, too stubborn and too hard to control." A woman's group in Kenya closely allied with the government accused her of violating African tradition because she was not docile and did not submit to men. They also accused her of having raised her voice against men in government, including the president. These criticisms revealed her formidable commitment and her fearlessness.

Kenya imports all of its oil, electricity and coal, so most of the people depend on local firewood for fuel. The country's forests have fallen at such a rate that its forest cover is now a fraction of what it once was. Most villagers suffer chronic wood shortages, and as Maathai pointed out:

> Poverty and need have a very close relationship with a degraded environment…
> … when you talk about the problems, you tend to disempower people. You tend to make people feel that there is nothing they can do, that they are doomed, that there is no hope. I realized that to break the cycle, one has to start with a positive step, and I thought that planting a tree is very simple, very easy— something positive that anybody can do.

Maathai noticed that many children in Kenya were being fed processed white bread, margarine and sweetened tea, a diet that kept them sickly and malnourished. She realized that this poor diet was a direct consequence of the scarcity of firewood for cooking. She made a connection between the large numbers of children in poor health and the degraded land resulting from the clearing of forests for plantations and fuel. Fields were heavily cultivated, and with no trees as protection against sun and rain, the soil baked and washed away. Maathai planted seven trees in her backyard and in 1976 began to show people how to plant trees in what would become known as the Green Belt Movement.

Maathai began to urge Kenyan farmers, 70 per cent of whom were women, to plant trees as a protective belt around fields. She toured schools to urge students to get their parents to plant native species of trees—baobabs, acacias, pawpaws, croton, cedars, citrus trees and figs. Recruiting children and women, Maathai's Green Belt Movement has grown into a major force in the country. By 2021, it had planted more than 51 million trees in Kenya. In 1986, Kenyan Green Belt members began to teach their methods to people from different African countries, an initiative that became the Pan-African Green Belt Movement, which includes active groups in Tanzania, Uganda, Malawi, Lesotho, Ethiopia and Zimbabwe.

The Green Belt Movement provides employment for people with disabilities, the poor and unemployed youth. It reaches many poor and illiterate women, gaining their trust and empowering them. The organization teaches them about proper nutrition and traditional foods and promotes family planning. In rural areas, staff of the movement distribute tools and train people to collect and plant seeds. It has become a major force for change in Kenya and was recognized by the United Nations as a model of grassroots conservation.

In 2002, Maathai was elected by a landslide and sent to parliament, where she worked tirelessly to raise awareness of environmental issues and to have the terrible burden of international debts cancelled for poor African countries. In 2004, she was awarded the Nobel Peace Prize for her work with the Green Belt Movement.

Karl-Henrik Robèrt and the Natural Step

Over billions of years a toxic stew of inorganic compounds has been transformed by cells into mineral deposits, forests, fish, soil, breathable air and water—the foundation of our economy and of our healthy existence. With sunlight as the sole energy supply those natural resources have been created in growing, self-sustaining cycles—the "waste" from one species providing nutrition for another. The only processes that we can rely on indefinitely are cyclical; all linear processes must eventually come to an end.

—KARL-HENRIK ROBÈRT, "Educating a Nation: The Natural Step"

The story of Karl-Henrik Robèrt is like a Hollywood movie. Robèrt was a renowned scientist and doctor specializing in cancer in children. Over the years he began to realize that many of the children who came to him had forms of cancer that were probably induced by environmental agents. He was amazed by the love that parents devoted to their children. "They would do anything within their ability for those children," he told me, but their concerns for the environment, which may have been the source of their children's problems, were not nearly as intense. Robèrt pondered this and finally drafted a document that was his attempt to define the conditions necessary to have a healthy and sustainable society. In it he warns that

up to now, much of the debate over the environment has had the character of monkey chatter amongst the withering leaves of a dying tree—the leaves representing specific, isolated problems... very few of us have been paying attention to the environment's trunk and branches. They are deteriorating as a result of pro-cesses about which there is little or no controversy; and the thousands of individual problems that are the subject of so much debate are, in fact, manifestations of systemic errors that are undermining the foundations of society.

In his analysis, Robèrt realized that in nature the utilization of energy and resources is circular—one species' waste product is another's opportunity. Human beings have changed that by breaking the circles and creating linear models of production in which matèrial is taken from the Earth, manufactured or used, and then ultimately discarded as waste or garbage. Robèrt tried to write out the principles of sustainability.

What Robèrt did then is remarkable. He sent his document to fifty of Sweden's leading scientists and asked them to critique his ideas. He received a flood of critical comments, but instead of being defeated by it, he incorporated their suggestions into a redraft and sent it back again, inviting further comment. He did this twenty-one times until he had honed the ideas into four fun-damental concepts with which none of the fifty scientists could disagree. Robèrt had scientific consensus!

He then took the document to the king of Sweden and asked the king to endorse the document—and he did. Robèrt went to Swedish television and asked for airtime to discuss the ideas—and he got it. He went to star Swedish performers and asked them to help in a program—and they did. Eventually, he sent a printed explanation and audiotape of his ideas to every household and every school in Sweden—4.3 million copies.

Robèrt's ideas have become known as the Natural Step and are being taught in all schools in Sweden. They have also been adopted as standards by more than fifty corporations in that country. Now the Natural Step has spread to several other coun-tries around the world. What are the four "systems conditions" that scientists agree are essential to sustainable societies? Although deceptively simple, the principles represent a profound change in how we perceive our activities on the planet:

- Nature cannot withstand a systematic buildup of dispersed matter mined from the Earth's crust (minerals, oil, etc.).
- Nature cannot withstand a systematic buildup of persistent compounds made by humans (e.g., PCBs).
- Nature cannot withstand a systematic deterioration of its capacity for renewal (e.g., harvesting fish faster than they can be replenished, converting fertile land to desert).
- Therefore, if we want life to continue, we must (a) be efficient in our use of resources and (b) promote justice—because ignoring poverty will lead the poor, for short-term survival, to destroy resources (e.g., the rainforests) that we all need for long-term survival.

We live in two interpenetrating worlds. The first is the living world, which has been forged in an evolutionary crucible over a period of four billion years. The second is the world of roads and cities, farms and artifacts, that people have been designing for themselves over the last few millennia. The condition that threatens both worlds—unsustainability—results from a lack of integration between them.

—SIM VAN DER RYN and STUART COWAN, *Ecological Design*

Robèrt's ideas are taking root in many companies and countries. Communities such as Whistler, B.C., and Canmore, Alberta, are undergoing explosive growth and have been working with the Natural Step Canada to find ways to be sustainable, and large corporations such as Interface and IKEA have adopted the Natural Step to reduce their ecological impact. By completely rethinking the way they do business and changing processes including purchase of materials, manufacturing, transportation, construction of facilities, maintenance and waste management, corporations reduce waste, become more efficient and increase profitability.

The Natural Step takes the Earth's absorptive and regenerative capacity as the standard against which human activity is weighed, thereby relieving human experts from trying to set limits based on incomplete science. In accepting the system's conditions, companies and groups must find their own solutions and ways to try to live within them, thereby claiming the process as their own. It is an exciting, positive process that is having a huge impact.

I don't believe that the solutions in society will come from the left or the right or the north or the south. They will come from... islands of people with integrity who want to do something... This is what a network should do—identify the people who would like to do something good. And they are everywhere. This is how the change will appear—you won't notice the difference. It won't be anyone winning over anyone. It will just spread... It will be something that the intelligent people do...
　—KARL-HENRIK ROBÈRT, "That Was When I Became a Slave"

Defender of Biodiversity: Vandana Shiva

The Earth family has been not just all humans of diverse societies, but all beings. The mountains and rivers are beings too. In Hindi, the words Vasudhaiva Kutumbam mean "Earth Family," the democracy of all life, all the little beings and the big ones with no hierarchy because you have no idea ecologically how things fit in the web of life.
　—VANDANA SHIVA, quoted in K. Ausubel, *Restoring the Earth*

Vandana Shiva is an Indian with a passion and commitment engendered by her upbringing. Her mother was a senior education officer who, after India was partitioned from Pakistan, became a farmer. Vandana remembers that her mother often pointed out that the forest is a model of existence. Vandana's father was a senior forest conservationist who often took her with him on trips to the woods in the foothills of the Himalayas. In fact, she never saw a city until she was fifteen years old, and by then she had developed a deep bond with nature that underlies everything she does and thinks.

It was her love of nature that led Shiva to pursue graduate studies in Canada in quantum physics, the study of the fundamental structure of all matter. This field taught her about unpredictability and diversity in the most elementary particles in the universe, concepts that suffuse her current ideas and actions. While at university, she would spend part of her summers trekking in areas her father had travelled, and it was on one of these trips that she encountered women in the famous Chipko (literally, "tree huggers") movement, in which women hugged trees to prevent them from being cut down. She began to work with the movement.

By the time she graduated with a PhD, she had turned down positions in North America to return to India, just as the vaunted benefits of the Green Revolution were being questioned. As part of the Green Revolution, in many countries traditional, locally adapted, genetically diverse grain seeds were replaced by a few highly selected strains that required great amounts of fertilizers, water, herbicides (to reduce plant competition) and insecticides, as well as agricultural machinery to plant and harvest the new grains. In a country such as India, where at the time nearly 70 per cent of the population was involved in peasant agriculture, the impact of the Green Revolution was immense. Self-sufficient communities that had grown produce for local needs were transformed into suppliers for foreign markets, highly dependent on expertise from elsewhere and locked into an expensive kind of agriculture that demanded use of more machinery. According to Shiva:

The modern world has built its ideas of nature and culture on the model of the industrial factory—judging a forest, for example, on the worth of its timber rather than its life support capacity.

Shiva saw that the effect of the Green Revolution had been to deplete diversity, which is the critical shield against pest outbreaks. Diversity was necessary to enrich the soil, to provide food for livestock and to feed the people. So in the end, the soil is degraded, communities suffer, and only the companies make money. Shiva began to work with farmers to fight against the transnational seed companies and to preserve the genetic diversity that was perpetuated by traditional agriculture. In 1991, she wrote *The Violence of the Green Revolution* to counter the claimed benefits.

Shiva also founded the Research Foundation for Science, Technology and Natural Resources Policy. Through the foundation, she discovered, among other things, that companies in India have obtained rulings that enabled them to patent hybrid seeds as "inventions," thereby gaining complete control of the very basis of agriculture. Once patented, a seed cannot be reused, traded or resold without payment to the patent holder. Furthermore, by manipulating one or two genes out of perhaps thirty thousand to a hundred thousand genes in a plant, a company could claim it as a new life form. When she realized the implications of what was happening, Shiva decided to fight corporate control over seeds:

If for the textile revolution, Gandhi picked up the spinning wheel, then it makes sense in the period of the engineering of life to pick up the seed... Each bit of seed tells the story of the community. In the seed is a political statement of what kind of life we want, what kind of agriculture we want, what relationship with the soil we want.

As a quantum physicist, Shiva knows that everywhere there is *potential* and that there is always a variety of possible outcomes. Freedom for us is keeping the potentials or options open, not restricting the possibilities. When large organizations control the means of production and spread uniform techniques or ideas to homogenize production and impose certainty, our options are reduced and we become vulnerable to unforeseen events.

When the $47 billion company Cargill, the largest grain trader in the world, filed for patents involving traditional Indian seeds, Shiva knew the company had to be confronted. If patents were issued, farmers could ultimately be prevented from using their own seeds. Indian farmers considered seeds to be their intellectual property and were outraged at Cargill's actions. In 1993, farmers invaded Cargill administrative offices in Delhi and burned their files. Months later, farmers razed Cargill's $2.5 million seed-processing facility, and soon after that, over half a million farmers rallied in Bangalore to protest plant-patenting laws. After the mass demonstration, Cargill soon announced it would no longer patent seeds in India.

To Shiva, biotechnology and the patenting of life forms are a grotesque extension of resourcism, a mindset that looks out at the world and evaluates everything in it according to potential utility and control:

> When life has become a mine for genes, nothing is safe. That's really the basis for our resistance in India to the slave trade of the twentieth century, the trade in genetic materials.

Shiva understands one of biology's most important lessons, that the key to life's resilience and to the farmer's crop is diversity:

The obligation to cultivate diversity is a cosmic obligation to keep a larger balance in place. That's the way in which we have been undertaking seed conservation in India.

She has created ten community seed banks to preserve the diversity that exists in the farmers' local grain varieties. She knows that over time these banks will become a priceless legacy for future generations to ward off the uncertainties of a changing world. She wants patent-free zones to be established in India and is working for a global network of farmers and gardeners cooperating in the preservation of seed stocks. At the same time, she works with the Third World Network, based in Malaysia, to keep abreast of the rapid changes in biotechnology and to fight the profit-driven agendas of transnational corporations and economic organizations such as the World Trade Organization.

Muhammad Yunus and the Grameen Bank

The myth that credit is the privilege of a few fortunate people needs to be exploded. You look at the tiniest village and the tiniest person in that village: a very capable person, a very intelligent person. You have only to create the proper environment to support these people so that they can change their own lives.

—MUHAMMAD YUNUS, quoted in D. Borstein,
"The Barefoot Bank With Cheek"

For people who have not been able to find steady employment or who do not own a car or a home, obtaining a bank loan is out of the question. Without collateral or a steady job, such people appear to be poor risks for borrowing money. It is hard enough for a person in a country such as Australia, Canada or the United States, so imagine what it must be like for someone in one of the poorest nations on Earth. Only a few years ago, Bangladesh was regarded as an economic basket case, and some people were suggesting that aid to the country should be halted because the population was beyond resuscitation. Well, Muhammad Yunus put the lie to that notion.

Yunus had worked on a PhD in economics in the United States in the 1960s, and the student activism of that time inflamed his idealism. Returning to Bangladesh in 1972, Yunus was keen to help improve the lot of his fellow citizens. In 1974, during a terrible famine in Bangladesh, Yunus was inspired to fund a small group of families so that they could create small items for sale. He believed that making such loans could alleviate widespread rural poverty. He took his students to meet local farmers and villagers with a view to helping them. In 1976, while he was head of the Department of Economics at the University of Chittagong, Yunus was walking in a village when he had an epiphany in an encounter with Sufiya Khatun, a widow. She was weaving bamboo stools for sale, and Yunus was shocked to learn that her profits were a mere two cents a day. When he asked her why she made so little, Khatun told him that she had to buy her bamboo in cash and the only person who would lend her money charged so much that she barely made a profit.

Yunus decided to find out whether there were others in the village with the same problem, and with his students, he compiled a list of forty-two people, who all together needed $26 to buy materials and to work freely. Yunus says:

> I felt extremely ashamed of myself being part of a society that could not provide twenty-six dollars to forty-two able, skilled human beings who were trying to make a living.

For loans of such tiny amounts, the clerical paperwork in regular banks would cost more than the loans. Besides, the people were extremely poor and illiterate, and they lacked collateral.

Two years later, Yunus set up the first branch of the Grameen (meaning "village") Bank. This bank was different from others in that (1) loans were to be repaid on time, (2) only the poorest—the landless—were eligible for loans, and (3) women, who occupied the lowest social and economic rung, would be sought out for loans. Loans are made to women whose family assets fall far below a regular bank's requirements. Instead of providing collateral, each woman becomes a member of a five-person group that assumes the collective debt of all five members. In turn, they become part of a forty-member centre, which meets weekly.

The bank has been a spectacular success. By 1983, the Grameen Bank had 86 branches serving 58,000 clients, and by May 2006 more than 2185 branch offices served 69,140 villages and 6.39 million customers, 96 per cent of whom were women. Remarkably, 98.45 per cent of its loans are repaid on time. As of November 2019, the Grameen Bank has provided services to over 9.6 million members in 81,678 villages—more than 93 per cent of the villages in Bangladesh. Of close to 30 million Grameen borrowers in Bangladesh, more than half have risen out of acute poverty, which is measured by such criteria as having all children of school age attending school, all household members eating three meals a day, a sanitary toilet, a rainproof house, clean drinking water and the ability to repay a US$8-a-week loan. For his groundbreaking work, Yunus was awarded the 2006 Nobel Peace Prize.

More than four hundred different kinds of businesses have been started with Grameen Bank loans—for example, businesses that husk rice, make ice cream sticks and process mustard oil. Today the example of the Grameen Bank has inspired similar banks in many parts of the world, including the wealthiest nation on Earth, the United States. *Microeconomics, microloans* and *microbanks* are terms that have even come to the attention of the World Bank and the United Nations. It is a fitting tribute to the work of Muhammad Yunus.

Making a Difference

These are just a few of the literally hundreds of eco-heroes who have already had a significant effect on their societies and beyond. Joining them are millions of people donating money and selflessly working without public recognition as volunteers, carrying out tasks from writing letters to staffing blockades and joining public protests. There is joy in the companionship of others working to make a difference for future generations, and there is hope. Each of us has the ability to act powerfully for change; together we can regain that ancient and sustaining harmony, in which human needs and the needs of all our companions on the planet are held in balance with the sacred, self-renewing processes of Earth.

The natural world is subject as well as object. The natural world is the maternal source of our being as earthlings and life-giving nourishment of our physical, emotional, aesthetic, moral and religious existence. The natural world is the larger sacred community to which we belong. To be alienated from this community is to become destitute in all that makes us human. To damage this community is to diminish our own existence.

—THOMAS BERRY, *The Dream of the Earth*

Afterword

A S I WRITE THIS, there are, in the words of the *Washington
Post*, "no fewer than five powerful heat domes swelling over
the landmasses of the Northern Hemisphere." A week or two
ago, the temperature in David Suzuki's British Columbia reached
49.6°C (121°F), probably the highest temperature Canada has seen
in millions of years. Today, the smoke from fires in Canada and
the American West has degraded the air quality on the East
Coast enough that we're told to refrain from outdoor exercise,
and huge new wildfires have broken out in Siberia. Meanwhile,
as central Europe tries to clean up after record flooding, China
is reporting... record flooding: Zhengzhou, home base for making iPhones, got 635 millimetres (25 inches) of rain in twenty-four
hours—literally a year's worth of rain in a day. There are all-time
record temperatures in Turkey and in Ireland.

All of which is to say, it's a hard backdrop against which
to read *The Sacred Balance*. Hard because it was written in the
moment when the author was still, mostly, pointing to the damage that was coming, offering a warning and a set of truly useful
prescriptions. We didn't heed that warning, and we didn't follow
those prescriptions—at least until COVID struck, we kept using
more fossil fuel year after year, and watching as the temperature

climbed. Even now, even as we get pummelled day after day with more fires, more floods, more storms, more droughts, our leaders continue to insist that we must dig up more stuff to burn.

Sometimes I think of the limits of persuasion: David Suzuki is year after year voted the most-admired Canadian; there can hardly be a Canadian who hasn't listened to him speak. And yet Justin Trudeau could go to Texas a few years ago and say, to a cheering crowd of oil industry executives, "no country would find 173 billion barrels of oil in the ground and leave them there." Certainly not Canada, which has literally socialized a pipeline so it can keep the most carbon-intensive oil on earth flowing, even in the face of protest from First Nations, even though the last thing the sensitive and beautiful oceanscape off Suzuki's Vancouver needs is a sharp increase in tanker traffic. It's as if the words have just bounced off the eyes and ears of people.

But maybe not, or not entirely. Because in the generation since David Suzuki first gave us the great gift of this book, something else has happened: the rise of the largest global movement ever, the one determined to hold industry to account and change the dynamic of our decline. And his words have had much to do with it—they've reached enough eyes and ears (enough hearts, really) to help build that groundswell. I've gotten to watch as people have started flooding into streets and jails, trying and succeeding in the effort to block pipelines, or to block the flow of money to Big Oil. Yes, there's a hideous lag effect—the damage we didn't prevent two decades ago is what's driving today's destruction. Yes, all we can do is keep from making it worse than it would be. But that's enough for many of us.

And that "us" is very different than it was. David tells the story, early in these pages, of how a visit with the Haida, as they fought clearcutting, helped shake his assumptions about the world; his foundation has gone on to do remarkable work with Indigenous activists around the world. Their ascension to the forefront of the movement against planetary destruction has been strategically important (from the tar sands to Standing Rock to Burnaby to the coalfields of Australia, Indigenous lands lie atop big deposits of carbon, or athwart the routes necessary to get it to market), but even more spiritually important. Indeed, Suzuki's early life as a bench scientist and his later life as an advocate alongside First Nations people, strikes me as the very exemplar of one of the most important facts about our time: the most ancient wisdom traditions and the most modern find themselves in sync. That is, the view from the satellite and the supercomputer, and the view from the sweat lodge, seem to me to mesh—and what they agree on is that the conventional wisdom of the world (especially that we can just keep growing our economies forever) is dangerously wrong.

That convergence is powerful, and in the decade ahead it will help make real the final challenge to the corporate might that threatens to overwhelm us. It will be a remarkable battle. But it does come very late. To read the remarkable words in this volume is to be reminded exactly how much warning we've had, how little we've done and what valuable things we have lost as a result.

Please don't let us lose any more.

—BILL MCKIBBEN

Notes

The numbers on the left
refer to page numbers.

CHAPTER 1

21 Robin Wall Kimmerer, *Braiding Sweetgrass: Indigenous Wisdom, Scientific Knowledge, and the Teachings of Plants* (Minneapolis: Milkweed Editions, 2013).

21 Size of brain, Carl Sagan, *Broca's Brain* (New York: Random House, 1974).

22 Brain's need to create order, François Jacob, *The Logic of Living Systems: A History of Heredity* (London: Allen Lane, 1970).

22 Synapses per neuron, Alison Gopnik, Andrew N. Meltzoff and Patricia K. Kuhl, *The Scientist in the Crib: What Early Learning Tells Us About the Mind* (New York: HarperCollins, 1999).

24 Stephen L. Talbott, "The Language of Nature," *The New Atlantis*, Winter 2007, https://www.thenewatlantis.com/publications/the-language-of-nature/.

25 Maria Montessori, *To Educate the Human Potential* (Madras: Kalakshetra, 1948).

26 John Donne, "The First Anniversary," *The Poems of John Donne* (New York: Oxford University Press, 1957).

26 Bernard Lown and Evjueni Chazov, quoted in P. Crean and P. Kome, eds.,

Peace, a Dream Unfolding (Toronto: Lester & Orpen Dennys, 1986).

27 Charles R. Darwin, *On the Origin of Species* (London: John Murray, 1859).

27 Gerardo Reichel-Dolmatoff, *Amazonian Cosmos: The Sexual and Religious Symbolism of the Tukano Indians* (Chicago: University of Chicago Press, 1971).

28 Stephen Jay Gould, *Wonderful Life: The Burgess Shale and the Nature of History* (New York: W.W. Norton, 1989).

28 Donald R. Griffin, *Animal Thinking* (Cambridge, MA: Harvard University Press, 1984).

29 Santiago Ramón y Cajal, *Recollections of My Life* (Cambridge, MA: Harvard University Press, 1969).

30 Roger Sperry, "Changed Concepts of Brain and Consciousness: Some Value Implications," *Zygon: Journal of Religion and Science* 20, no. 1 (1985): 41–57.

31 Stuart Kauffman, *At Home in the Universe: The Search for the Laws of Self-Organization and Complexity* (New York: Oxford University Press, 1995).

32 Ian Lowe, personal communication.

33 Millennium Ecosystem Assessment, *Ecosystems and Human Well-Being: Biodiversity Synthesis* (Washington, D.C.: World Resources Institute, 2005).

33 Robin Wall Kimmerer, *Gathering Moss: A Natural and Cultural History of Mosses* (Corvallis: Oregon State University Press, 2003).

33 Richard Cannings and Sydney Cannings, *British Columbia: A Natural History* (Vancouver: Greystone Books, 1996).

33 Nathan Cobb, *Nematodes and Their Relationships* (Washington, D.C.: Department of Agriculture, 1915), http://www.ars.usda.gov/Main/docs .htm?docid=9626. Accessed March 29, 2007.

33 Census of Marine Life, http://www .coml.org. Accessed June 15, 2006.

35 Ronald W. Clark, *Einstein: The Life and Times* (New York: Avon Books, 1971).

35 Jonathan Marks, *Human Biodiversity: Genes, Race and History* (Hawthorne, N.Y.: Aldine de Gruyter, 1995).

36 U.S. Department of Energy Office of Science, Office of Biological and Environmental Research, "Human Genome Project," http://www.ornl .gov/sci/technresources/Human _Genome/home.shtml. Accessed June 15, 2006.

37 The Chimpanzee Sequencing and Analysis Consortium, "Initial Sequence of the Chimpanzee Genome and Comparison With the Human Genome," *Nature* 437 (2005): 69–87.

37 Eric Lander, in "Episode 1: Journey Into New Worlds," *The Sacred Balance* (Toronto: CBC, 2003).

38 Brian Goodwin, quoted in John Brockman, "A New Science of Qualities: A Talk With Brian Goodwin," *Edge* 15, April 29, 1997.

39 Brian Swimme, *The Hidden Heart of the Cosmos*, 1996. Video available from Centre for the Story of the Universe, Mill Valley, California.

39 Robert Browning, "Caliban Upon Setebos," in *The Norton Anthology of English Literature* (New York: W.W. Norton, 1987).

39 Simon Nelson Patten, quoted in H. Allen, "Bye-Bye America's Pie," *Washington Post*, February 11, 1992.

39 Paul Wachtel, *The Poverty of Affluence: A Psychological Portrait of the American Way of Life* (Gabriola Island, B.C.: New Society, 1988).

40 Consumption as the answer, Alan Thein Durning, *How Much Is Enough? The Consumer Society and the Future of the Earth* (New York: W.W. Norton, 1992).

40 Victor Lebow, quoted in Vance Packard, *The Waste Makers* (New York: David McKay, 1960).

40 American economy's "ultimate purpose," R. Reich, *The Work of Nations: Preparing Ourselves for 21st-Century Capitalism* (New York: Alfred A. Knopf, 1991).

40 Mr. Jean Chrétien, quoted in Jill Vardy and Chris Wattie, "Shopping Is Patriotic, Leaders Say," *National Post*, September 28, 2001.

41 White House press release, "At O'Hare, President Says 'Get on Board,'" September 27, 2001, http://www.whitehouse.gov/news /releases/2001/09/20010927-1.html. Accessed June 25, 2006.

41 Donald R. Keough, quoted in R. Cohen, "For Coke, World Is Its Oyster," *New York Times*, November 21, 1991.

41 Campaign for a Commercial-Free Childhood, "Marketing to Children: An Overview," http://www.com

mercialfreechildhood.org. Accessed June 25, 2006.

42 Juliet Schor, *Born to Buy: The Commercialized Child and the New Consumer Culture* (New York: Scribner, 2004).

42 Robin Wall Kimmerer, "Speaking of Nature," *Orion Magazine*, March/April 2017, https://orionmagazine.org/article/speaking-of-nature/.

42 P.M. McCann, K. Fullgrabe and W. Godfrey-Smith, *Social Implications of Technological Change* (Canberra: Department of Science and Technology, 1984).

43 Allen D. Kanner and Mary E. Gomes, "The All-Consuming Self," *Adbusters*, Summer 1995.

44 José Lutzenberger, "Conference on Global Environment," April 30, 1990, C-SPAN, video recording, 34:33, https://www.c-span.org/video/?12101-1/conference-global-environment.

44 Benjamin Franklin, quoted in H. Goldberg and R.T. Lewis, *Money Madness: The Psychology of Saving, Spending, Loving, and Hating Money* (New York: William Morrow, 1978).

45 Leah Watkins, Ryan Gage, Moira Smith, Christina McKerchar, Robert Aitken, and Louise Signal, "An Objective Assessment of Children's Exposure to Brand Marketing in New Zealand (Kids'Cam): A Cross-Sectional Study," *Lancet* 6, no. 2 (February 2022): E132–38, https://doi.org/10.1016/S2542-5196(21)00290-4.

45 Peter Doran, *Doing More With Less: Ensuring Sustainable Consumption and Production* (IISD, August 2021), https://www.iisd.org/system/files/2021-08/still-one-earth-sustainable-consumption.pdf.

45 Statista Research Department, *Median Size of Single Family Housing Unit in the United States From 2000 to 2020*, March 28, 2022, https://www.statista.com/statistics/456925/median-size-of-single-family-home-usa/.

45 Simon Torkington, "The Pandemic Has Changed Consumer Behaviour Forever—and Online Shopping Looks Set to Stay," World Economic Forum, July 7, 2021, https://www.weforum.org/agenda/2021/07/global-consumer-behaviour-trends-online-shopping.

45 Laura Cozzi and Apostolos Petropoulos, "Carbon Emissions Fell Across All Sectors in 2020 Except for One—SUVs," International Energy Agency, January 15, 2021, https://www.iea.org/commentaries/carbon-emissions-fell-across-all-sectors-in-2020-except-for-one-suvs.

45 Patrick Greenfield, "What Is Deforestation—and Is Stopping It Really Possible?," *Guardian*, November 1, 2021, https://www.theguardian.com/environment/2021/nov/01/cop26-what-is-deforestation-and-is-stopping-it-really-possible-aoe.

47 Richard Louv, *Last Child in the Woods: Saving Our Children From Nature-Deficit Disorder* (Chapel Hill: Algonquin Books, 2005).

48 George Eliot, *Middlemarch* (London: Penguin Books, 1871).

48 Albert Einstein, quoted in P. Crean and P. Kome, eds., *Peace, a Dream Unfolding* (Toronto: Lester & Orpen Dennys, 1986).

49 World Commission on Environment and Development, *Our Common*

Future (New York: Oxford University Press, 1987).

50   Kimmerer, *Braiding Sweetgrass.*

51   Carl Sagan et al., "Preserving and Cherishing the Earth: An Appeal for Joint Commitment in Science and Religion," quoted in Peter Knudtson and David T. Suzuki, *Wisdom of the Elders* (Toronto: Stoddart, 1992).

52   Lyall Watson, *Supernature* (Anchor Press, 1973).

CHAPTER 2

53   Harlow Shapley, *Beyond the Observatory* (New York: Scribners, 1967).

53   Gerard Manley Hopkins, "The Blessed Virgin Compared to the Air We Breathe," in *Poems of Gerard Manley Hopkins* (Oxford: Oxford University Press, 1948).

54   Plato, *Phaedro*, quoted in D.I. Blumenstock, *The Ocean of Air* (New Brunswick, N.J.: Rutgers University Press, 1959).

54   Julie Payette, in "Episode 1: Journey Into New Worlds," *The Sacred Balance* (Toronto: CBC, 2003).

56   Father José de Acosta, *Natural and Moral History* (1590), quoted in Blumenstock, *The Ocean of Air.*

57   Built-in safety measures, A. Despopoulos and S. Silbernagl, *Color Atlas of Physiology*, 4th ed. (New York: Theime Medical Publishers, 1991).

57   Jonathan Weiner, *The Next One Hundred Years* (New York: Bantam Books, 1990).

63   Shapley, *Beyond the Observatory.*

65   Cynthia Beall et al., "Ventilation and Hypoxic Ventilatory Response of Tibetan and Aymara High Altitude Natives," *American Journal of Physical Anthropology* 104, no. 4 (1997): 427–47.

65   Cynthia Beall, "Andean, Tibetan and Ethiopian Patterns of Adaptation to High-Altitude Hypoxia," *Integrative and Comparative Biology* 46, no. 1 (2006): 18–24.

67   Composition of primordial atmosphere, James Lovelock, *Gaia: The Practical Science of Planetary Medicine* (London: Allen & Unwin, 1991).

69   Lynn Margulis, quoted in "Gaia Is a Tough Bitch," in John Brockman, *The Third Culture* (New York: Simon & Schuster, 1995).

73   Pressure of lower atmosphere, Blumenstock, *The Ocean of Air.*

78   Jules M. Blais, David W. Schindler et al., "Accumulation of Persistent Organochlorine Compounds in Mountains of Western Canada," *Nature* 395 (1998): 585–88.

78   David Schindler, in "Episode 2: The Matrix of Life," *The Sacred Balance* (Toronto: CBC, 2003).

79   Eric Dewailly, quoted in Marla Cone, *Silent Snow: The Slow Poisoning of the Arctic* (New York: Grove Press, 2005).

80   Fred Hoyle, quoted in E. Goldsmith, P. Bunyard, N. Hildyard and P. McCully, *Imperiled Planet* (Cambridge, MA: MIT Press, 1990).

82   V. Shatalov, quoted in K.W. Kelley, ed., *The Home Planet* (Herts, U.K.: Queen Anne Press, 1988).

CHAPTER 3

85   Psalm 104:10–11, Authorized (King James) Version.

85   Amount of water on Earth, M. Keating, *To the Last Drop: Canada and the*

World's Water Crisis (Toronto: Macmillan Canada, 1986).

86 William Shakespeare, *The Tempest*, in *The Works of Shakespeare* (New York: Macmillan, 1900).

87 Mark A. McMenamin and Dianna L.S. McMenamin, *Hypersea: Life on Land* (New York: Columbia University Press, 1994).

92 700 tonnes of water moving through a tree, Michael Keller, in "Episode 2: The Matrix of Life," *The Sacred Balance* (Toronto: CBC, 2003).

92 Jack Vallentyne, *American Society of Landscape Architects* (Ontario Chapter) 4, no. 4 (September–October 1987).

94 Capturing of hydrogen, James Lovelock, *Gaia: The Practical Science of Planetary Medicine* (London: Allen & Unwin, 1991).

94 Vladimir Vernadsky, quoted in M.I. Budyko, S.F. Lemeshko and V.G. Yanuta, *The Evolution of the Biosphere* (Dordrecht: D. Reidel Publishing, 1986).

96 Daniel Hillel, *Out of the Earth: Civilization and the Life of the Soil* (Herts, U.K.: Maxwell Macmillan, 1986).

96 Intake of water balanced with daily losses, A. Despopoulos and S. Silbernagl, *Color Atlas of Physiology*, 4th ed. (New York: Thieme Medical Publishers, 1991).

99 Peter Warshall, "The Morality of Molecular Water," *Whole Earth Review*, Spring 1995.

101 Seth Shostak, "8 Worlds Where Life Might Exist," SETI Institute, March 23, 2006, http://www.space .com/search forlife/060323_seti

_biomes.html. Accessed July 5, 2006.

103 Richard Saykally, in "Episode 2: The Matrix of Life," *The Sacred Balance* (Toronto: CBC, 2003).

105 Samuel Taylor Coleridge, *The Rime of the Ancient Mariner*, in *The Norton Anthology of English Literature* (New York: W.W. Norton, 1987).

105 Fresh water as rarest form of water on Earth, W.E. Akin, *Global Patterns in Climate, Vegetation, and Soils* (Norman: University of Oklahoma Press, 1991).

107 Genesis 2:10, Revised Standard Version.

107 Leonardo da Vinci, quoted in Hillel, *Out of the Earth.*

108 Great Lakes as containing 20 per cent of all fresh water on Earth, K. Lanz, *The Greenpeace Book of Water* (Newton Abbot, U.K.: David & Charles, 1995).

108 Volume and use of flowing river water, M. Keating, *To the Last Drop.*

108 Samuel Taylor Coleridge, "Kubla Khan," in *The Norton Anthology of English Literature* (New York: W.W. Norton, 1987).

109 Ecclesiastes 1:7, Revised Standard Version.

112 World Health Organization, "Drinking-Water," March 21, 2022, https://www.who.int/news-room /fact-sheets/detail/drinking-water.

112 World Water Council, "Water Crisis," http://www.worldwatercouncil.org /index.php?id=25. Accessed July 10, 2006.

112 Fred Pearce, "The Parched Planet," *New Scientist* 189 (2006): 32–36.

113 John G. Rodwan Jr., "Bottled Water 2019: Slower but Notable Growth,"

BWR, July/August 2020, https://
bottledwater.org/wp-content
/uploads/2021/03/2019BWstats
_BMcarticle_BWR_JulyAug2020
_lowres.pdf.

113 Natural Resources Defense Council,
"Bottled Water: Pure Drink or Pure
Hype?," http://www.nrdc.org/water
/drinking/bw/bwinx.asp. Accessed
June 25, 2006.

113 Consumer Reports, "What's in That
Bottle?," http://www.consumer
reports.org/cro/food/drinkingwater
-safety-103/whats-in-bottled-water
/index.htm. Accessed March 29,
2007.

115 Theo Colborn, Dianne Dumanoski
and John Peterson Myers, *Our Stolen
Future* (New York: Dutton Books,
1996).

116 Rachel Carson, *Silent Spring* (Boston:
Houghton Mifflin, 1962).

CHAPTER 4

118 Genesis 3:19, Revised Standard
Version.

118 Quote from origin story, Genesis
2:7–8, 15, Revised Standard Version.

119 Daniel Hillel, *Out of the Earth: Civilization and the Life of the Soil* (Herts, U.K.:
Maxwell Macmillan, 1991).

120 Luther Standing Bear, *My People the
Sioux*, ed. E.A. Brininstool (reprint,
Lincoln: University of Nebraska
Press, 1975).

121 Aldo Leopold, *A Sand County
Almanac* (New York: Oxford
University Press, 1949).

121 Philip Cohen, "Clay's Matchmaking
Could Have Sparked Life," *New
Scientist* 23 (2003), http://www.new
scientist.com/article.ns?id=dn4307.
Accessed July 15, 2006.

121 Philip Ball, "Shaped From Clay: Minerals Help Molecules Thought to
Have Been Essential for Early Life to
Form," *Nature*, November 3,
2005, http://www.news.nature
.com/news/2005/051031-10.html.
Accessed July 15, 2006.

122 Gisday Wa and Delgam Uukw,
The Spirit in the Land (Gabriola, B.C.:
Reflections, 1987).

122 S. Lomayaktewa, M. Lansa,
N. Nayatewa, C. Kewanyama,
J. Pongayesvia, T. Banyaca Sr.,
D. Monogyre and C. Shattuck,
mimeographed statement of Hopi
religious leaders, 1990, quoted in
Peter Knudtson and David T. Suzuki,
Wisdom of the Elders (Toronto:
Stoddart, 1992).

123 Carl Sagan et al., "Preserving and
Cherishing the Earth: An Appeal for
Joint Commitment in Science and
Religion," quoted in Knudtson and
Suzuki, *Wisdom of the Elders*.

124 Yvonne Baskin, *Under Ground: How
Creatures of Mud and Dirt Shape Our
World* (Washington, D.C.: Islands
Press, 2005).

124 Leonardo da Vinci, quoted in Hillel,
Out of the Earth.

124 Elaine Ingham, *The Soil Foodweb: Its
Importance in Ecosystem Health*, http://
www.rain.org/~sals/ingham.html.
Accessed July 8, 2006.

125 Elaine Ingham, in "Episode 1: Journey
Into New Worlds," *The Sacred Balance*
(Toronto: CBC, 2003).

126 Tullis Onstott, in "Episode 2: The
Matrix of Life," *The Sacred Balance*
(Toronto: CBC, 2003).

128 Hillel, *Out of the Earth*.

132 Homer, quoted in R.S. Gottlieb,
ed., *This Sacred Earth: Religion,*

Nature, Environment (London: Routledge, 1996).

136 S.W. Simard and D.M. Durall, "Mycorrhizal Networks: A Review of Their Extent, Function and Importance," *Canadian Journal of Botany* 82 (2004): 1140–65.

136 N.A. Campbell et al., *Biology* (San Francisco: Benjamin Cummings, 1999).

140 Horizons of soil, Frank Press and Raymond Siever, *Earth* (San Francisco: W.H. Freeman & Company, 1982).

141 Deuteronomy 8:8–9, Revised Standard Edition.

141 Need for vitamins, A. Despopoulos and S. Silbernagl, *Color Atlas of Physiology*, 4th ed. (New York: Theime Medical Publishers, 1991).

146 Gaia Vince, "Your Amazing Regenerating Body," *New Scientist* 190 (2006): 2556.

149 Vernon Gill Carter and Tom Dale, *Topsoil and Civilization* (Norman: University of Oklahoma Press, 1974).

149 Senator Herbert Sparrow, *Soil at Risk: Canada's Eroding Future* (Ottawa: Government of Canada, 1984).

149 Depletion of topsoil in U.S., D. Helms and S.L. Flader, eds., *The History of Soil and Water Conservation* (Berkeley: University of California Press, 1985).

149 Depletion of topsoil in Australia, *Australia: State of the Environment* (Victoria: CSIRO, 1996).

149 Quote by Anonymous, unattributed source, in Carter and Dale, *Topsoil and Civilization*.

150 David Pimentel, "Constraints on the Expansion of Global Food Supply," *Ambio* 23 (1994).

150 D. Pimentel, "Soil as an Endangered Ecosystem," *BioScience* 50 (2000): 947.

150 D. Pimentel and M. Pimentel, "World Population, Food, Natural Resources, and Survival," *World Futures* 59 (2003): 145–67.

150 Use of controlled burning by Aborigines, T. Flannery, *The Future Eaters* (Victoria, Australia: Reed Books, 1994).

151 Deborah Bird Rose, *Nourishing Terrains: Australian Aboriginal Views of Landscape and Wilderness* (Parkes: Australian Heritage Commission, 1996).

152 Bernard Campbell, *Human Ecology* (New York: Heinemann Educational, 1983).

153 W.C. Lowdermilk, "Conquest of the Land Through 7,000 Years," Soil Conservation Service, Bulletin 99 (Washington, D.C.: U.S. Department of Agriculture, 1953).

153 Quote about the Waswanipi people, Knudtson and Suzuki, *Wisdom of the Elders*.

154 Marq de Villiers, *Water* (Toronto: Stoddart, 1999).

155 P.P. Micklin, "Desiccation of the Aral Sea: A Water Management Disaster in the Soviet Union," *Science* 241 (1998): 1170–76.

155 United States Agency for International Development, "USAID Environmental Restoration of the Aral Sea," https://www.usaid.gov/central-asia-regional/fact-sheets/usaid-environmental-restoration-aral-sea. Accessed April 29, 2022.

155 Deutsche Presse-Agentur, "Water to Dust: As the Aral Sea Dries Up, Some Battle for Its Survival," *The Star*, January 24, 2022,

https://www.thestar.com.my
/lifestyle/living/2022/01/24/water
-to-dust-as-the-aral-sea-dries
-up-some-battle-for-its-survival.

158 Miguel Altieri, *Agroecology: The Science of Sustainable Agriculture*, 2nd ed. (Boulder, CO: Westview, 1995).

158 Heather Archibald, "Organic Farming: The Trend is Growing!," *Canadian Agriculture at a Glance*, Statistics Canada Catalogue No. 96-325-XPB (Ottawa: Statistics Canada, 1999).

159 Leopold, *A Sand County Almanac*.

CHAPTER 5

160 Hildegarde of Bingen, quoted in David Maclagan, *Creation Myths: Man's Introduction to the World* (London: Thames & Hudson, 1977).

160 Wallace Stevens, "Sunday Morning," in *The Norton Anthology of Poetry* (New York: W.W. Norton, 1923).

160 Rig-Veda X.129, quoted in Maclagan, *Creation Myths*.

162 The body as a house with air-conditioning and heating systems, A. Despopoulos and S. Silbernagl, *Color Atlas of Physiology*, 4th ed. (New York: Theime Medical Publishers, 1991).

167 Myth of Prometheus, *Larousse Encyclopedia of Mythology* (London: Batchworth Press, 1959).

172 Stanley L. Miller, "A Production of Amino Acids Under Possible Primitive Earth Conditions," *Science* 117 (1953): 528–29.

172 Experiments generating molecules needed for all macromolecules, C. Ponnamperuma, *The Origins of Life* (New York: Dutton, 1972).

177 A. Melis, L. Zhang et al., "Sustained Photobiological Hydrogen Gas Production Upon Reversible Inactivation of Oxygen Evolution in the Green Alga *Chlamydomonas reinhardtii*," *Journal of Plant Physiology* 122 (2000): 127–36.

177 American Society of Plant Biologists, "Scientists Use Algae to Find Valuable New Source of Fuel," http://www.aspb.org/publicaffairs /news/melis.cfm. Accessed August 10, 2006.

178 Fossil fuels as a once-only gift, E.J. Tarbuck and F.K. Lutgens, *The Earth: An Introduction to Physical Geology* (Columbus: Merrill Publishing, 1987).

179 Colin Campbell, quoted in John Vidal, "The End of Oil Is Closer Than You Think," *Guardian*, April 21, 2005, http://www.guardian.co.uk/life /featurestory/0,13026,1464050,00 .html. Accessed March 29, 2007.

179 Jeffrey Dukes, "Burning Buried Sunshine: Human Consumption of Ancient Solar Energy," *Climate Change* 61 (2003): 31–44.

180 Malcolm Smith on seeing pictures of Earth at night, quoted in the *Guardian*, April 25, 1989.

181 Maurice Strong, quoted in the *Guardian*, April 25, 1989.

181 Naomi Oreskes, "Beyond the Ivory Tower: The Scientific Consensus on Climate Change," *Science* 306 (2004): 1686.

181 Intergovernmental Panel on Climate Change, *Global Warming of 1.5°C* (Geneva: IPCC, 2018).

182 Fred Pearce, "Climate Change: Menace or Myth?," *New Scientist* 185 (2005): 38–43.

182 Amory Lovins, "More Profit With Less Carbon," *Scientific American* 293 (2005): 74–83.

182 Nicholas Stern, *The Economics of Climate Change: The Stern Review* (Cambridge: Cambridge University Press, 2007).

183 McKinsey & Company, *The Net-Zero Transition: What It Would Cost, What It Could Bring*, January 2022, https://www.mckinsey.com/business-functions/sustainability/our-insights/the-net-zero-transition-what-it-would-cost-what-it-could-bring.

183 Ian Parry, Simon Black and Nate Vernon, *Still Not Getting Energy Prices Right: A Global and Country Update of Fossil Fuel Subsidies*, International Monetary Fund working paper, September 24, 2021, https://www.imf.org/en/Publications/wp/Issues/2021/09/23/Still-Not-Getting-Energy-Prices-Right-A-Global-and-Country-Update-of-Fossil-Fuel-Subsidies-466004.

183 Intergovernmental Panel on Climate Change, IPCC press release 2022/15/PR, April 4, 2022, https://www.ipcc.ch/report/ar6/wg3/resources/press/press-release/.

185 Outlining an economy, David Pimentel, "Natural Resources and an Optimum Human Population," *Population and Environment* 15, no. 5 (1994):

186 Buying time for the design and construction of living spaces, M. Safdie with W. Kohn, *The City After the Automobile* (Toronto: Stoddart, 1997).

186 Manufacturing processes reducing use of energy and materials, E. von Weizsacker, A. Lovins and L.H. Lovins, *Factor 4: Doubling Wealth—Halving Resource Use* (London: Allen & Unwin, 1997).

CHAPTER 6

187 Charles R. Darwin, *On the Origin of Species* (London: John Murray, 1859).

187 Lynn Margulis, *Five Kingdoms* (San Francisco: W.H. Freeman & Company, 1982).

187 Black Elk, quoted in T.C. McLuhan, *Touch the Earth* (New York: Promontory Press, 1971).

189 Aimée Craft, *Treaty Words: For as Long as the Rivers Flow* (Toronto: Annick Press, 2021).

190 Percentage of all species that are now extinct, R. Leakey and R. Lewin, *The Sixth Extinction: Biodiversity and Its Survival* (London: Weidenfeld & Nicolson, 1995).

190 Bepkororoti, Kayapo leader in Brazil, quoted in Oxfam report, "Amazonian Oxfam's Work in the Amazon Basin."

191 Sunderlal Bahuguna, spokesperson for Chipko movement, quoted in E. Goldsmith, P. Bunyard, N. Hildyard and P. McCully, *Imperiled Planet* (Cambridge, MA: MIT Press, 1990).

193 Biomass of micro-organisms compared with biomass of larger organisms, Stephen Jay Gould, *Full House: The Spread of Excellence From Plato to Darwin* (New York: Crown, 1996).

194 Stephen Jay Gould, "Planet of the Bacteria," *Washington Post Horizon* 119 (1996): H1.

195 Richard Strohman, "Crisis Position," *Safe Food News* 2000, http://www .mindfully.org/GE/Strohman-Safe -Food.htm. Accessed March 29, 2007.

196 Victor B. Scheffer, *Spires of Form: Glimpses of Evolution* (Seattle: University of Washington Press, 1983).

200 Catherine Larrère, "A Necessary Partnership With Nature," UNESCO *Courier,* May 2000.

200 Dennis Banks, quoted in "Ojibwe Warrior, American Indian Movement Co-Founder and Civil Rights Leader Dennis Banks Has Died," Native American Netroots, November 18, 2019, http://native americannetroots.net/diary /Ojibwe-Warrior-American-Indian -Movement-Co-Founder-And-Civil -Rights-Leader-Dennis-Banks-Has -Died/.

201 Black Elk, quoted in J.G. Neihardt, *Black Elk Speaks* (New York: Washington Square Press, 1959).

201 Stephen Jay Gould, *The Mismeasure of Man* (New York: W.W. Norton, 1981).

203 Edward O. Wilson, *The Diversity of Life* (Cambridge, MA: Harvard University Press, 1992).

204 Losses from southern corn blight, Gail Schumann, *Plant Diseases: Their Biology and Social Impact* (Saint Paul, MN: American Phytopathological Society, 1991).

204 Peter Wohlleben, *The Hidden Life of Trees: What They Feel, How They Communicate—Discoveries From a Secret World,* trans. Jane Billinghurst (Vancouver: Greystone Books, 2016).

205 Introduced species in tropical forests, Francis Hallé, personal communication.

208 Vandana Shiva, *Monocultures of the Mind: Perspectives on Biodiversity and Biotechnology* (New York: Oxford University Press, 1993).

210 Lynn Margulis, "Symbiosis and Evolution," *Scientific American* 225 (1971): 48–57.

211 Lynn Margulis, *Symbiotic Planet: A New Look at Evolution* (New York: Basic Books, 1998).

212 Edward O. Wilson, in "Learning to Love the Creepy Crawlies," *The Nature of Things* (Toronto: CBC, 1996).

213 Howard T. Odum, *Environment, Power and Society* (New York: John Wiley & Sons, 1971).

215 Jonathan Weiner, *The Next One Hundred Years* (New York: Bantam Books, 1990).

216 Nourishment by nature, Y. Baskin, *The Work of Nature: How the Diversity of Life Sustains Us* (Washington, D.C.: Island Press, 1997).

217 James Lovelock and Gaia, ibid.

218 Human harnessing of primary productivity of the planet, P.M. Vitousek, P.R. Ehrlich, A.H. Ehrlich and P.A. Matson, "Human Appropriation of the Products of Photosynthesis," *BioScience* 36 (1986): 368–73.

218 Alan Thein Durning, "Saving the Forests: What Will It Take?" Worldwatch Paper 117, December 1993.

220 John Fowles, quoted in McLuhan, *Touch the Earth.*

221 Edward O. Wilson, "Biophilia and the Conservation Ethic," in *The Biophilia Hypothesis,* ed. S.R. Kellert and E.O. Wilson (Washington, D.C.: Island Press, 1993).

221 E. Goldsmith, P. Bunyard, N. Hildyard and P. McCully, *Imperiled*

Planet (Cambridge, MA: MIT Press, 1990).

221 Wilson, "Biophilia and the Conservation Ethic."

222 Intergovernmental Science-Policy Platform on Biodiversity and Ecosystem Services, *Summary for Policymakers of the Global Assessment Report on Biodiversity and Ecosystem Services of the Intergovernmental Science-Policy Platform on Biodiversity and Ecosystem Services*, 2019, https://ipbes.net/sites/default/files/ipbes_7_10_add.1_en_1.pdf.

222 World Wildlife Fund, *Living Planet Report 2020*, https://f.hubspotusercontent20.net/hubfs/4783129/LPR/PDFS/ENGLISH-FULL.pdf.

222 "UN Report: Nature's Dangerous Decline 'Unprecedented'; Species Extinction Rates 'Accelerating,'" United Nations Sustainable Development Goals, May 6, 2019, https://www.un.org/sustainabledevelopment/blog/2019/05/nature-decline-unprecedented-report/.

222 David L. Wagner, Eliza M. Grames, Matthew L. Forister, May R. Berenbaum and David Stopak, "Insect Decline in the Anthropocene: Death by a Thousand Cuts," *Proceedings of the National Academy of Sciences* 118, no. 2 (2021), https://doi.org/10.1073/pnas.2023989118.

222 Millennium Ecosystem Assessment, *Ecosystems and Human Well-Being: Biodiversity Synthesis* (Washington, D.C.: World Resources Institute, 2005).

222 Ransom Myers and Boris Worm, "Rapid Worldwide Depletion of Predatory Fish Communities," *Nature*, May 15, 2003, 280.

223 Paul Crutzen and Eugene Stoermer, "Anthropocene," http://www.mpch-mainz.mpg.de/~air/anthropocene/Text.html. Accessed August 15, 2006.

223 John A. Livingston, *One Cosmic Instant* (Toronto: McClelland & Stewart, 1973).

224 Richard Preston, *The Hot Zone* (New York: Pantheon Books, 1994).

224 "The Problem," Preventing Pandemics at the Source, https://www.preventingfuturepandemics.org/the-problem. Accessed March 16, 2022.

225 Jonathan Schell, *The Abolition* (New York: Alfred A. Knopf, 1984).

227 Rachel Carson, *Silent Spring* (Boston: Houghton Mifflin, 1962).

228 Dam in Tasmania, "Pedder 2000: A Symbol of Hope at the New Millennium," Global 500 Forum Newsletter no. 13, February 1995.

229 Pope John Paul II, "The Ecological Crisis: A Common Responsibility," message for celebration of World Day of Peace, 1990.

229 Wilson, "Biophilia and the Conservation Ethic."

230 St. Francis of Assisi, "The Canticle of Brother Sun," in E. Doyle, *St. Francis and the Song of Brotherhood* (New York: Seabury Press, 1980).

CHAPTER 7

231 Abraham H. Maslow, *Motivation and Personality* (New York: Harper & Row, 1970).

232 Ibid.

232 Ashley Montagu, *The Direction of Human Development* (New York: Harper & Brothers, 1955).

233 Ashley Montagu, *Growing Young* (New York: McGraw-Hill, 1981).

234 Desiderius Erasmus (1465–1536), quoted in P. Crean and P. Kome, eds., *Peace, a Dream Unfolding* (Toronto: Lester & Orpen Dennys, 1986).

235 Experiments with baby monkeys, H.F. Harlow and M.K. Harlow, "Social Deprivation in Monkeys," *Scientific American* 207 (1962): 136–46.

235 Montagu, *Direction of Human Development*.

236 Montagu, *Growing Young*.

237 Alfred Adler, *Social Interest: A Challenge to Mankind* (New York: Putnam, 1938).

237 Montagu, *Growing Young*.

238 Dr. Helen Fisher, quoted in Lauren Slater, "Love: The Chemical Reaction," *National Geographic*, February 2006, 32–49.

239 Society for Neuroscience, "Love and the Brain," *Brain Briefings*, December 2005, http://apu.sfn.org/index .cfm?pagename=brainBriefings _loveAndTheBrain. Accessed August 20, 2006.

239 Society for Neuroscience, "Scientists Uncover Neurobiological Basis for Romantic Love, Trust and Self," *ScienceDaily*, November 11, 2003, http://www.sciencedaily.com /releases/2003/11/031111064658.htm. Accessed August 21, 2006.

240 Correlation between happiness and other factors, D.G. Myers and E. Diener, "The Pursuit of Happiness," *Scientific American*, May 1996.

241 John Donne, *Devotions Upon Emergent Occasions*, Meditation XVII, in *Complete Poetry and Selected Prose* (New York: Random House, 1929).

241 Montagu, *Growing Young*.

242 Ibid.

243 André Gide, *The Journals of André Gide*, vol. 1 (New York: Alfred A. Knopf, 1947).

244 Liz Warwick, "Dr. Michael Meaney: More Cuddles, Less Stress!," *Bulletin of the Centre of Excellence for Early Childhood Development* 4, no. 2 (2005): 2.

245 J.G. Ruiz-Peláez, N. Charpak and L.G. Cuervo, "Kangaroo Mother Care, an Example to Follow From Developing Countries," *British Medical Journal* 329 (2004): 1179–81.

246 Debora MacKenzie, "Hamster Dads Make Wonderful Midwives," *New Scientist* 166 (2000): 13.

246 K.E. Wynne-Edwards and S.J. Berg, "Changes in Testosterone, Cortisol and Estradiol Levels in Men Becoming Fathers," *Mayo Clinic Proceedings* 76 (2001): 582–92.

246 Maslow, *Motivation and Personality*.

247 Conditions in the *leagane*, Children's Health Care Collaborative Study Group, "Romanian Health and Social Care System for Children and Families: Future Directions in Health Care Reform," *British Medical Journal* 304 (1992): 556–59.

248 Quote about need for adult engagement with infants, S. Blakeslee, "Making Baby Smart: Words Are Way," *International Herald Tribune*, April 18, 1997.

248 Study about death of children in Romania, D.R. Rosenberg, K. Pajer and M. Rancurello, "Neuropsychiatric Assessment of Orphans in One Romanian Orphanage for 'Unsalvageables,'" *Journal of the American Medical Association* 268 (1992): 3489–90.

249 Quotes about state of Romanian children adopted by Americans, Johnson et al., "The Health of Children Adopted From Romania," *JAMA* 268, no. 24 (1992): 3446–51.

249 Quote about response of children to improved conditions, ibid.

249 Sarah Jay, "When Children Adopted Overseas Come With Too Many Problems," *New York Times*, June 23, 1996.

249 Quote about Victor Groze's findings, ibid.

250 Paroma Basu, "Psychologists Glimpse Biological Imprint of Childhood Neglect," University of Wisconsin–Madison press release, 2005, http://www.eurekalert.org/pub_releases/2005-11/uow-pgb111705.php. Accessed August 20, 2006.

250 Anthony Walsh, *The Science of Love* (Buffalo: Prometheus Books, 1991).

250 Diane Luckow, "Tracking the Progress of Romanian Orphans," SFU Public Affairs and Media Relations, press release 28 (2003): 1.

250 Elinor Ames, *The Development of Romanian Children Adopted Into Canada: Final Report* (Burnaby: Simon Fraser University, 1997).

251 Percentage of first grade–aged children separated from mothers, I. Zivcic, "Emotional Reactions of Children to War Stress in Croatia," *Journal of the American Academy of Child & Adolescent Psychiatry* 32 (1993): 709–13.

251 Symptoms of children separated from families, L.C. Terr, "Childhood Traumas: An Outline and Overview," *American Journal of Psychiatry* 148 (1991): 10–20.

251 Children's need to trust a close adult, Zivcic, "Emotional Reactions of Children."

252 Montagu, *Direction of Human Development*.

254 A.L. Engh, J.C. Beehner et al., "Behavioral and Hormonal Response to Predation in Female Chacma Baboons (*Papio hamadryas ursinus*)," *Proceedings of the Royal Society B: Biological Sciences* 273 (2006): 707–12.

254 Greg Lester, "Baboons in Mourning Seek Comfort Among Friends," University of Pennsylvania press release, 2006.

254 Frans de Waal, *Good Natured* (Cambridge, MA: Harvard University Press, 1996).

254 Rossella Lorenzi, "Elephants Mourn Their Dead," *News in Science*, November 4, 2005.

254 J. Gamble, "Humor in Apes," *Humor* 14, no. 2 (2001): 163–79.

255 John Robinson and Caroline Van Bers, *Living Within Our Means* (Vancouver: David Suzuki Foundation, 1996).

255 Anthony Stevens, "A Basic Need," *Resurgence*, no. 174 (January/February 1996).

256 Prince Modupe, *I Was a Savage* (London: Museum Press, 1958).

257 Results of putting economic goals above social goals, J.P. Grayson, "The Closure of a Factory and Its Impact on Health," *International Journal of Health Sciences* 15 (1985): 69–93.

257 Results of putting economic goals above social goals, R. Catalano, "The Health Effects of Economic Insecurity," *American Journal of Public Health* 81 (1991): 1148–52.

257 Results of putting economic goals above social goals, S. Platt, "Unemployment and Suicidal Behaviour: A Review of the Literature," *Society Science Medicine* 19 (1984): 93–115.

257 Results of putting economic goals above social goals, L. Taitz, J. King, J. Nicholson and M. Kessel, "Unemployment and Child Abuse," *British Medical Journal Clinical Research Edition* 294 (1987): 1074–76.

257 Results of putting economic goals above social goals, M. Brenner, "Economic Change, Alcohol Consumption and Heart Disease Mortality in Nine Industrialized Countries, *Social Science Medicine* 25 (1987): 119–32.

258 Need for meaningful employment, R.L. Jin, C.P. Shah and T.J. Svoboda, "The Impact of Unemployment on Health: A Review of the Evidence," *Canadian Medical Association Journal* 153 (1995): 529–40.

258 Need for meaningful employment, J.L. Brown and E. Pollitt, "Malnutrition, Poverty and Intellectual Development," *Scientific American*, February 1996, 38–43.

258 Mahatma Gandhi, quoted in P. Crean and P. Kome, eds., *Peace, a Dream Unfolding* (Toronto: Lester & Orpen Dennys, 1986).

259 Ivaluardjuk, quoted in K. Rasmussen, "Intellectual Culture of the Caribou Eskimoes," *Report of the Fifth Thule Expedition, 1921–1924*, vol. 7, 1930.

260 Definition of "biophilia," Edward O. Wilson, *Biophilia: The Human Bond With Other Species* (Cambridge, MA: Harvard University Press, 1984).

260 Henry David Thoreau, *Walden* (Princeton: Princeton University Press, 1971).

260 Wilson, *Biophilia*.

261 Howard Frumkin, "Beyond Toxicity: Human Health and the Natural Environment," *American Journal of Preventive Medicine* 20 (2001): 234–40.

261 Francesca Lyman, "The Geography of Health," *Land & People*, Fall 2002.

261 Robin Marwick, "Therapy and Service Dogs: Friends and Healers," *AboutKidsHealth*, 2006.

261 Oliver Sacks, *A Leg to Stand On* (London: Gerald Duckworth & Co. Ltd., 1984).

262 Studies supporting biophilia hypothesis, *The Biophilia Hypothesis*, ed. S.R. Kellert and E.O. Wilson (Washington, D.C.: Island Press, 1993).

263 Ibid.

264 Anita Barrows, "The Ecological Self in Childhood," *Ecopsychology Newsletter*, no. 4 (Fall 1995).

265 Rising number of children with asthma, "The Scary Spread of Asthma and How to Protect Your Kids," *Newsweek*, May 26, 1997.

265 Vine Deloria, *We Talk, You Listen* (New York: Delta Books, 1970).

266 Paul Shepard, *Nature and Madness* (San Francisco: Sierra Club Books, 1982).

267 Carin Gorrell, "Nature's Path to Inner Peace," *Psychology Today* 34 (2001): 62.

267 Sarah Conn, in "Episode 4: Coming Home," *The Sacred Balance* (Toronto: CBC, 2003).

CHAPTER 8

268 W.B. Yeats, "Sailing to Byzantium," in *The Norton Anthology of English*

Literature (New York: W.W. Norton, 1927).

269 Hopi myth, B.C. Sproul, *Primal Myths: Creation Myths Around the World* (New York: Harper & Row, 1979).

270 African myths, ibid.

270 T.S. Eliot, *The Waste Land*, in *The Norton Anthology of English Literature* (New York: W.W. Norton, 1922).

271 Ker Than, "Why Great Minds Can't Grasp Consciousness," *Live Science*, http://www.livescience.com /humanbiology/050808_human _consciousness.html. Accessed September 1, 2006.

271 David J. Chalmers, "Facing Up to the Problem of Consciousness," *Journal of Consciousness Studies* 2 (1995): 200–219.

272 Kenneth Marten and Suchi Psarakos, "Evidence of Self-Awareness in the Bottlenose Dolphin (*Tursiops truncatus*)," in *Self-Awareness in Animals and Humans: Developmental Perspectives* (New York: Cambridge University Press, 1995).

272 Gordon Gallup, James Anderson and Daniel Shillito, "The Mirror Test," in *The Cognitive Animal: Empirical and Theoretical Perspectives on Animal Cognition* (Cambridge, MA: MIT Press, 2002).

273 Aztec lamentation, in Margot Astrov, ed., *American Indian Prose and Poetry* (New York: Capricorn, 1962).

275 Tomson Highway, quoted in Suzanne Methot, "The Universe of Tomson Highway," *Quill & Quire*, November 1998, https:// quillandquire.com/authors /the-universe-of-tomson -highway/.

275 M.K. Dudley, in *This Sacred Earth: Religion, Nature, Environment*, ed. R.S. Gottlieb (London: Routledge, 1996).

276 Quotes about Aborigine beliefs, D. Kinsley, *Ecology and Religion: Ecological Spirituality in Cross-Cultural Perspective* (New York: Prentice Hall, 1995).

277 Robin Dunbar, "We Believe," *New Scientist* 189 (2006): 28–33.

278 Paul Bloom, "Is God an Accident?," *Atlantic Monthly*, http://www.the atlantic.com/doc/200512/god -accident. Accessed September 10, 2006.

278 Paul Bloom, "Natural-Born Dualists," *Edge: The Third Culture*, May 13, 2004, http://www.edge.org/3rd _culture/bloom04/bloom04_index .html. Accessed September 9, 2006.

279 Lao Tzu, *The Complete Works of Lao Tzu: Tao Te Ching and Hua Hu Ching*, trans. Hua-Ching Ni (Santa Monica, CA: SevenStar Communications, 1979).

279 Annie Dillard, *Teaching a Stone to Talk* (New York: HarperCollins, 1982).

281 Daniel Swartz, "Jews, Jewish Texts and Nature," in *This Sacred Earth: Religion, Nature, Environment*, ed. R.S. Gottlieb (London: Routledge, 1996).

282 Richard Wilbur, "Epistemology," in *The Norton Anthology of Modern Poetry*, ed. R. Ellmann and R. O'Clair (New York: W.W. Norton, 1973).

282 Neil Evernden, *The Natural Alien: Humankind and Environment* (Toronto: University of Toronto Press, 1993).

283 Jacques Monod, *Chance and Necessity* (New York: Vintage Books, 1972).

285 Lame Deer, quoted in D.M. Levin, *The Body's Recollection of Being: Phenomenal Psychology and the Deconstruction of Nihilism* (London: Routledge & Kegan Paul, 1985).

285 Erich Neumann, *The Origins and History of Consciousness* (Princeton: Princeton University Press, 1954).

286 Henri Bergson, quoted in Levin, *The Body's Recollection of Being.*

286 Joseph Meeker, *Minding the Earth* (Alameda, CA: Latham Foundation, 1988).

287 William Blake, "Auguries of Innocence," in *Complete Writings* (New York: Oxford University Press, 1972).

287 Thomas Hardy, "Transformations" (New York: Penguin Books, 1960).

289 Evernden, *The Natural Alien.*

289 W.B. Yeats, "Among School Children," in *The Norton Anthology of English Literature* (New York: W.W. Norton, 1927).

290 Psalm 96:11–13, Revised Standard Version.

291 Quote about mortals leaping and dancing, Levin, *The Body's Recollection of Being.*

291 Genesis 2:19, Revised Standard Version.

292 Clearcut as meadow, P. Moore, *Pacific Spirit* (West Vancouver: Terra Bella, 1996).

292 Seamus Heaney, "The First Words," in *The Spirit Level* (London: Faber & Faber, 1996).

293 William Wordsworth, "Lines Composed a Few Miles Above Tintern Abbey, on Revisiting the Wye During a Tour," in *The Norton Anthology of English Literature* (New York: W.W. Norton, 1987).

293 Quotes from artists, B. and T. Roszak, "Deep Form in Art and Nature," *Resurgence*, no. 176 (May/June 1996).

297 Pablo Neruda, quoted in R.S. Gottlieb, ed., *This Sacred Earth.*

CHAPTER 9

298 Richard Wagamese, quoted in "As It Happened: The Archive Edition— Richard Wagamese on his 2014 Novel Medicine Walk," *As It Happens,* CBC Radio, August 30, 2019, https://www.cbc.ca/radio/asithappens/as-it-happens-friday-edition-1.5260084/as-it-happened-the-archive-edition-richard-wagamese-on-his-2014-novel-medicine-walk-1.5230835/.

302 Paul Ehrlich and Anne Ehrlich, *Betrayal of Science and Reason: How Anti-Environmental Rhetoric Threatens Our Future* (Washington, D.C.: Island Press, 1997).

312 Goethe, quoted in P. Crean and P. Kome, eds., *Peace, a Dream Unfolding* (Toronto: Lester & Orpen Dennys, 1986).

313 Al Gore, *Earth in the Balance* (Boston: Houghton Mifflin, 1992).

313 Number of government institutions dealing with environment, M.K. Tolba, "Redefining UNEP," *Our Planet* 8, no. 5 (1997): 9–11.

314 Isaiah 11:6, Revised Standard Version.

315 Severn Cullis-Suzuki, *Tell the World* (Toronto: Doubleday, 1993).

317 Greta Thunberg, "'Our House Is on Fire," speech at the 2019 World Economic Forum in Davos, Switzerland, *Guardian*, January 25, 2019, https://

www.theguardian.com
/environment/2019/jan/25/our
-house-is-on-fire-greta-thunberg16
-urges-leaders-to-act-on-climate.

319 Section on Ian Kiernan based on
an interview with David Suzuki in
Sydney, April 22, 1997.

320 Section on William McDonough
based on K. Ausubel, *Restoring
the Earth: Visionary Solutions From
the Bioneers* (Tiburon, CA: H.J.
Kramer, 1997).

321 William McDonough and Michael
Braungart, *Cradle to Cradle: Remaking
the Way We Make Things* (New York:
North Point Press, 2002).

324 Wangari Maathai, quoted in P. Sears,
"Wangari Maathai: 'You Strike the
Woman...,'" *In Context* 28 (Spring
1991).

325 Wangari Maathai, quoted in Aubrey
Wallace, *Eco-Heroes: Twelve Tales of
Environmental Victory* (San Francisco:
Mercury House, 1993).

326 Wangari Maathai, *The Green Belt
Movement: Sharing the Approach
and the Experience* (New York:
Lantern Books, 2004).

327 Karl-Henrik Robèrt, "Educating
a Nation: The Natural Step," *In
Context* 28 (Spring 1991).

329 Sim Van der Ryn and Stuart Cowan,
Ecological Design (Washington, D.C.:
Island Press, 1996).

329 Karl-Henrik Robèrt, *The Natural
Step Story: Seeding a Quiet Revolution*
(Gabriola Island, B.C.: New Society
Publishers, 2002).

330 Karl-Henrik Robèrt, "That Was
When I Became a Slave," excerpts
from an interview by Robert
Gilman and Nikolaus Wyss, *In
Context* 28 (Spring 1991).

330 Section on Vandana Shiva based on
K. Ausubel, *Restoring the Earth.*

332 Vandana Shiva, quoted in *E—The
Environmental Magazine*, January/
February, n.d.

334 Section on Muhammad Yunus
based on D. Borstein, "The Barefoot
Bank With Cheek," *Atlantic Monthly*,
December 1955.

335 Muhammad Yunus, *Banker to the
Poor: Micro-lending and the Battle
Against World Poverty* (New York:
Public Affairs, 1999).

335 Muhammad Yunus and Alan Jolis,
*Banker to the Poor: The Autobiography
of Muhammad Yunus, Founder of the
Grameen Bank* (London: Aurum
Press Ltd., 1998).

337 Thomas Berry, *The Dream of the
Earth* (San Francisco: Sierra Club
Books, 1988).

Index

Figures and tables indicated by page numbers in italics

Credits

The following sources have given permission for quoted material:

From *Biophilia* by E.O. Wilson. Copyright © 1984 by E.O. Wilson. Reprinted by permission of the author. From *The Diversity of Life* by E.O. Wilson. Copyright © 1992 by E.O. Wilson. Reprinted by permission of the author. From "Biophilia and the Conservation Ethic" by E.O. Wilson, in *The Biophilia Hypothesis*, edited by S.R. Kellert and E.O. Wilson. Copyright © 1993 by E.O. Wilson. Reprinted by permission of the author. From "Natural Resources and an Optimum Human Population" by D. Pimentel, in *Population and Environment*, vol. 15, no. 5. Copyright © 1994 by D. Pimentel. Reprinted by permission of the author. From *Creation Myths: Man's Introduction to the World* by D. Maclagan. Copyright © 1977 by D. Maclagan. Reprinted by permission of Thames & Hudson. From Worldwatch Paper 117, "Saving the Forests: What Will It Take?" by A.T. Durning, Worldwatch Institute, December 1993. Reprinted by permission of Worldwatch Institute. From *Restoring the Earth* © 1997 by Kenny Ausubel. Reprinted by permission of H.J. Kramer, P.O. Box 1082, Tiburon, CA. All rights reserved. From "All-Consuming Passion: Waking Up From the American Dream." Created by the New Road Map Foundation, P.O. Box 15981, Seattle, WA 98115. Reprinted by permission of the New Road Map Foundation. From *Living Within Our Means* by John Robinson and Caroline Van Bers. Copyright © 1996 by the David Suzuki Foundation. Reprinted by permission of the David Suzuki Foundation. From *Wisdom of the Elders* by Peter Knudtson and David Suzuki. Copyright © 1992 by Peter Knudtson and David Suzuki. Reprinted by permission of the authors. From *In Context* 28, "Educating a Nation: The Natural Step," an interview with Karl-Henrik Robèrt. Reprinted by permission of the Context Institute (P.O. Box 946, Langley, WA 98260; URL: http:// www.context.org; email: ci@context.org). From "World Scientists' Warning to Humanity." Reprinted by permission of the Union of Concerned Scientists. From *Amazonian Cosmos: The Sexual and Religious Symbolism of the Tukano Indians* by G. Reichel-Dolmatoff. Copyright © 1971 by G. Reichel-Dolmatoff. Reprinted by permission of the University of Chicago Press. From *Peace, a Dream Unfolding*, editors P. Kome and P. Crean, Lester & Orpen Dennys Ltd., an imprint of Key Porter Books, Toronto, 1986. Reprinted by permission of Key Porter Books. From *Motivation and Personality* by A.H. Maslow. Copyright © 1970 by A.H. Maslow. Reprinted by permission of Addison-Wesley Educational Publishers Inc. From *Our Common Future*, World Commission on Environment and Development. Copyright © 1987 by Oxford University Press. Reprinted by permission of Oxford University Press. From "A Basic Need" by A. Stevens, in *Resurgence* 174, January/February 1996.

Acknowledgements

I AM GRATEFUL to many people who generously responded to my appeal for information, including Dr. Brian Holl (University of British Columbia), Dr. Craig Russell (University of British Columbia), Dr. Darrin Lehman (University of British Columbia), Dr. Robert Jin (Richmond, B.C.), Dr. Bill Fyfe (University of Western Ontario) and Dr. Tony Bai (St. Paul's Hospital, Vancouver, B.C.).

I was astonished by the generosity of people willing to review my work on short notice. My thanks to the reviewers of the first edition: Dr. Digby McLaren (chapter 1), Dr. David Bates (chapter 2), Dr. Jack Vallentyne (chapter 3), Dr. Les Lavkulich (chapter 4), Dr. David Brooks (chapter 5), Dr. Charles Krebs (chapter 6), Dr. Janine Brody (chapter 7) and Rev. Peter Hamel (chapter 8). All opinions, of course, are mine.

Many thanks to the following volunteers at the David Suzuki Foundation, who helped me get the first edition off the ground: Gina Agelidis, Robin Bhattacharya, Dr. Leslie Cotter, Catherine Fitzpatrick, Anna Lemke, Nicole Rycroft, Cathy St. Germain and Nick Scapillati. Thanks to Caterina Geuer for all of her help and to her partner, Chris Knight, for bringing Harlow Shapley's wonderful quote to my attention. Thank you, Christian Jensen, for the important task of tracking down permissions.

I am deeply grateful to Jack Stoddart, who respected me and my work enough to release me from our long-standing publishing relationship so that I could publish this book as a David Suzuki Foundation publication through Greystone Books.

Nancy Flight has been a wonderful editor, cheerleader and taskmaster. Rob Sanders has shared many air molecules with me and captured the vision of the book to throw his enthusiasm behind it.

Eveline de la Giroday kept the office running smoothly so that I could finish the first edition while Elois Yaxley fulfilled the same role so that I could finish the second edition.

Tara Cullis assumed all of the burdens of fulfilling the needs of children, home, friends and the foundation so that I could do this book.

For assistance with the 2007 edition, I would like to thank Andy Torr for his excellent research support and the Arcangelo Rea Family Foundation for assistance in funding.

Thank you to Ian Hanington for his help with updates to the 25th Anniversary Edition. To Robin Wall Kimmerer, who wrote the foreword, and Bill McKibben, who wrote the afterword—I am honoured by your contributions to this new edition.

Authors

DAVID SUZUKI is an award-winning geneticist and environmentalist who co-founded the David Suzuki Foundation in 1990. He is the author of more than forty books and is familiar to audiences around the world as host of CBC's *The Nature of Things*. From 1969 to 2001, he was a faculty member at the University of British Columbia, and he is currently professor emeritus. He is widely recognized as a world leader in sustainable ecology and has received numerous awards for his work, including a UNESCO prize for science, a United Nations Environment Programme medal and the Right Livelihood Award. He is also a Companion of the Order of Canada. He has 29 honorary degrees from universities in Canada, the U.S. and Australia. For his support of Canada's Indigenous Peoples, Suzuki has been honoured with eight names and formal adoption by two First Nations. He lives in Vancouver, British Columbia.

AMANDA MCCONNELL has written more than one hundred documentary films, many of them for *The Nature of Things*. She has a PhD in English literature, and she writes and gardens in Toronto, Ontario.

ADRIENNE MASON is the author of numerous books for adults and children, including *The Nature of Spiders*, *The Green Classroom*, *Living Things* and *Oceans*. She is managing editor of the science magazine KNOW: *The Science Magazine for Curious Kids* and has been nominated four times for a Science in Society book award. She holds a bachelor of science degree in biology from the University of Victoria, and she lives in Tofino, British Columbia, with her husband and two daughters.

IAN HANINGTON is senior editor and writer at the David Suzuki Foundation and is co-author with David Suzuki of *Just Cool It!* and *Everything Under the Sun*. He has worked as a news reporter, opinion-column writer, critic and editor for a variety of newspapers and magazines and was editor of Canada's largest alternative newsweekly, the *Georgia Straight*. He lives in Vancouver, British Columbia.

DAVID SUZUKI INSTITUTE

THE DAVID SUZUKI INSTITUTE is a non-profit organization founded in 2010 to stimulate debate and action on environmental issues. The Institute and the David Suzuki Foundation both work to advance awareness of environmental issues important to all Canadians.

We invite you to support the activities of the Institute. For more information please contact us at:

David Suzuki Institute
219 – 2211 West 4th Avenue
Vancouver, BC, Canada V6K 4S2
info@davidsuzukiinstitute.org
604-742-2899
www.davidsuzukiinstitute.org

Cheques can be made payable to the David Suzuki Institute.